Developing U.S. Army Officers' Capabilities for Joint, Interagency, Intergovernmental, and Multinational Environments

M. Wade Markel, Henry A. Leonard, Charlotte Lynch,
Christina Panis, Peter Schirmer, Carra S. Sims

Prepared for the United States Army
Approved for public release; distribution unlimited

ARROYO CENTER

The research described in this report was sponsored by the United States Army under Contract No. W74V8H-06-C-0001.

Library of Congress Cataloging-in-Publication Data

Developing U.S. Army officers' capabilities for joint, interagency, intergovernmental, and multinational environments / M. Wade Markel ... [et al.].
p. cm.
Includes bibliographical references.
ISBN 978-0-8330-5031-1 (pbk. : alk. paper)
1. United States. Army—Officers—Training of. 2. Command of troops. 3. Leadership—United States. 4. Unified operations (Military science) 5. Operational art (Military science) I. Markel, M. Wade.

UB413.D48 2011
355.5'50973—dc22

2010051766

Published 2011 by the RAND Corporation
1776 Main Street, P.O. Box 2138, Santa Monica, CA 90407-2138
1200 South Hayes Street, Arlington, VA 22202-5050
4570 Fifth Avenue, Suite 600, Pittsburgh, PA 15213-2665
RAND URL: http://www.rand.org/
To order RAND documents or to obtain additional information, contact
Distribution Services: Telephone: (310) 451-7002;
Fax: (310) 451-6915; Email: order@rand.org

Preface

The current security environment has significantly increased the salience of capabilities provided by other services, other U.S. government agencies, and multinational partners to Army operations. The Department of Defense has long emphasized joint operations; emphasis on interagency and multinational collaboration is somewhat newer. In each of these domains, however, actual operational practice has tended to advance faster than institutional support. In Iraq and Afghanistan, for example, brigade commanders command sizable elements from other services, respond to multinational commands, and integrate their operations with the activities of other agencies, both governmental and nongovernmental. Also, the events surrounding Hurricane Katrina reinforced once again the need for military leaders to be fully prepared to coordinate and conduct civil support operations.

It is thus important to examine career patterns and professional development needs in the context of requirements for officers to operate in the more complex joint, interagency, intergovernmental and multinational (JIIM) circumstances of today and tomorrow. The U.S. Army Human Resources Command's Officer Personnel Management System (OPMS) Task Force asked RAND Arroyo Center to analyze both the supply and the demand aspects of the identification and development of key competencies for success in JIIM environments.

This monograph should be of interest to policymakers concerned with the professional development of military officers and of other national security professionals.

The research was sponsored by the Commanding General, U.S. Army Human Resources Command and conducted in RAND Arroyo Center's Manpower and Training Program. RAND Arroyo Center, part of the RAND Corporation, is a federally funded research and development center sponsored by the United States Army. Questions and comments regarding this research are welcome and should be directed to the leader of the research team, M. Wade Markel, mmarkel@rand.org.

The Project Unique Identification Code (PUIC) for the research that produced this product is DAPEM08795.

For more information on RAND Arroyo Center, contact the Director of Operations (telephone 310-393-0411, extension 6419; FAX 310-451-6952; email Marcy_Agmon@rand.org), or visit Arroyo's web site at http://www.rand.org/ard/.

Contents

Figures

Tables

Summary

Introduction

Law, policy, and, most importantly, ongoing operations require the Department of Defense and the Army to develop a cadre of officers skilled in the integration of joint, interagency, intergovernmental, and multinational (JIIM) capabilities into military operations. The Army is responsible for developing the specific officer management policies to provide enough officers with the right capabilities to meet this demand. Developing such policies requires an understanding of the knowledge, skills, and abilities appropriate to the JIIM domains, identification of experiences that develop these capabilities, and assessment of the feasibility of different career paths in developing the required degree of proficiency. The U.S. Army Human Resources Command's Officer Personnel Management System (OPMS) Task Force is responsible for developing officer management policy to meet the needs of the future Army and future joint force.

Consonant with that responsibility, the OPMS Task Force asked RAND Arroyo Center to identify and describe the knowledge, skills, and abilities required to integrate JIIM capabilities into full-spectrum operations and to develop a framework that will enable the Army to better track and manage the inventory of officers who possess these capabilities. Our research team combined job analysis with inventory modeling to determine what capabilities Army officers require in the JIIM domains, what jobs develop those capabilities, and whether the Army could produce enough officers with the right capabilities to meet anticipated future demand.

Initially, the study team considered answering these questions with regard to the entire officer population. This broad scope would have been appropriate for several reasons. First, the Army consists of its active component and two reserve components, the Army Reserve and the Army National Guard. The demands that current operations place on these three components are qualitatively similar. Second, it seems intuitively obvious that reserve component officers, particularly those in the National Guard, would bring more experience and thus perhaps better insight into the intergovernmental domain. Time and resources, however, limited us to concentrating on the active component, a scope we could adequately cover in reasonable depth. Thus, while we are reasonably confident that the knowledge, skills, and abilities we identified apply to officers of all components, more work remains to be done in identifying developmental opportunities and inventory requirements for the reserve components.

Research Approach

Four research tasks made up our research approach. First, the research team conducted job analyses to ascertain the knowledge, skills, and abilities required in the JIIM domains. Job analysis is a discipline within the field of industrial and organizational psychology that identifies the major tasks that make up a given job, and the knowledge, skills, and abilities that allow incumbents to perform those jobs. Interviews with Army officers who had been successful in various aspects of JIIM environments, as well as with individuals from other services, U.S. government agencies, and nations, formed the major part of our job analyses. The full range of potential knowledge, skills, and abilities are listed and defined in Appendix A. Second, the team surveyed Army assignment officers to identify the types of positions that developed the desired capabilities. Third, the team developed alternative developmental patterns based on what we learned from our interviews and focus groups, prior RAND research, and existing literature about the development of expertise. We looked at alternative patterns that would allow the Army either to distribute these capabilities as widely as possible throughout its senior leadership, or to develop a set of experts in these domains by concentrating repeated developmental experiences on a smaller set of officers. Fourth, the team modeled the potential inventory of such officers that the Army could produce under either developmental pattern in order to validate those patterns' feasibility and assess their suitability. As it turned out, the last two research tasks became inextricably intertwined in an exploration of expertise.

As we mentioned above, this study focused on active component officers. That focus was not wholly exclusive: our respondents included officers from the reserve components and their interlocutors, and the knowledge, skills, and abilities we identified apply broadly to all officers. Time and resources limited our investigation of developmental opportunities to the active component, however. While many of those positions have identical or at least similar counterparts in the reserve components, we did not assess the degree to which positions unique to the reserve components, e.g., those on state standing joint force headquarters, contribute to the development of knowledge, skills, and abilities relevant to the JIIM domains. Intuitively, it would seem that those positions would provide excellent developmental experience, particularly in the intergovernmental domain. More work remains to be done in this arena.

Findings

Our investigation uncovered few surprises. Our respondents affirmed the proposition that basic military, branch, and functional area expertise were essential to success in JIIM settings. Interpersonal skills and other integration skills tend to be of primary importance in JIIM environments, in which success usually requires voluntary collaboration between independent organizations that are frequently pursuing different agendas. These and other general findings are described in greater detail below.

- Successful performance in joint, interagency, or multinational contexts requires the application of highly developed functional expertise to novel situations. Service, branch, or functional area expertise formed the cornerstone of Army officers' utility in a JIIM context. Officers need to understand their specialty well enough to think beyond an Army context, however. A military policeman must be able to think and to act like a policeman; a military engineer must be able to perform as an engineer, and so on.

- The JIIM domains are qualitatively distinct, if overlapping. Simply put, success in each of the JIIM domains requires a different set of knowledge, skills, and abilities; proficiency in one JIIM domain does not completely translate to proficiency in another. To the extent that actual expertise is required, developing it requires focus and multiple experiences in different contexts.
- The strategic, operational, tactical, and institutional echelons require distinctly different knowledge, skills, and abilities. That is, jobs at these different echelons differ in kind, not just in degree. A combatant command staff is not just a bigger, more capable brigade staff.
- Broadening experiences contribute significantly to competence in the JIIM domains. For some, that broadening experience was service in the Balkans; for others, it was a tour on a higher-level staff; and for one, it was working with KATUSAs (Koreans Attached to the U.S. Army). What all these experiences had in common was that they forced officers to cope with an unfamiliar context, and that mission success depended on effective, entirely voluntary cooperation from other individuals and organizations.
- In the current operating environment, Army officers have significantly increased opportunity to gain experience in one or more JIIM domains. Our survey of assignment officers indicated that even service in "Army" positions, such as battalion or brigade commander, executive officer, or operations officer, provided significant experience in integrating joint and multinational capabilities. Officers who served on division and higher echelon staffs also accumulated significant interagency experience. Development required deployment, however. In a garrison setting, those same positions provided little opportunity for developing JIIM-relevant knowledge, skills, and abilities.

- It should be possible to develop and maintain enough officers with the required knowledge, skills, and abilities in the JIIM domains widely. Our modeling showed that if the Army were to adopt a broad approach to managing experiences, over two-thirds of lieutenant colonels would acquire some sort of JIIM experience by the time they either retire or become colonels, and all colonels will have accumulated a JIIM assignment sometime in their career. Alternatively, the Army could produce substantially fewer "experts" (i.e., by concentrating multiple experiences on a smaller number of officers). It is likely that the resulting inventory could still satisfy demand. Either approach requires deliberate and effective management.

The need for breadth may be our most consequential finding. Army officers and individuals from other services, agencies, and nations with whom they worked largely agreed that competence in Army officers' branch and functional areas was fundamental to success in any JIIM experience. The Army already spends considerable time and effort developing and refining that competence. There is very little spare time in a typical career available to develop breadth, especially if one were to add a requirement to accrue experience in a JIIM domain. For now, experience in current operations mitigates this tension significantly (Wong, 2004). Over the longer term, however, the Army might consider tilting the balance between depth and breadth of experience toward breadth. Such a shift might be appropriate if the current security environment, characterized by irregular challenges, stability operations, and the concomitant importance of nonmilitary instruments of national power persists. Alternatively, the Army might conclude that the average career is simply not long enough to accommodate all the experience an officer must accrue, and thus seek to lengthen careers for at least some officers. If breadth is valuable, however, the Army will need to incentivize it. To the extent that officers view opportunities as "not career enhancing," they will avoid them.

The small size of our sample populations, especially for our assignment officer survey, should inspire some caution in acting on these findings. They are nonetheless consistent with other studies and the findings of the current Joint Qualification Board. We are reasonably confident of their general validity.

There is considerable room for further research. There is probably more to be done to refine understanding of the knowledge, skills, and abilities needed in the various JIIM domains and in other emerging domains relevant to the conduct of full-spectrum operations. In particular, further study is required to better understand the unique developmental opportunities inherent in reserve component assignments, particularly in the intergovernmental domain. A large-scale quantitative study would allow the Army to align particular knowledge, skills, and abilities with JIIM domains and with echelons. Such a study should probably consider the reserve components as well. In the same vein, the Army might want to better understand the alignment of knowledge, skills, and abilities with particular echelons more generally. Our investigation indicated that individuals working at strategic, operational, tactical, and institutional echelons required different sets of knowledge, skills, and abilities, including but not limited to those capabilities in the JIIM domains. Finally, the Army could further investigate the question of expertise, including the degree of proficiency required to perform certain jobs and how that proficiency is developed.

Acknowledgments

We wish to thank our many respondents, on whose thoughtful and cogent observations this study rests. We greatly appreciate their candor, as well as their generosity with their time and energy. Special thanks are owed to those who helped us coordinate focus groups: Colonel Julie Manta of the U.S. Army War College, Lieutenant Colonel Justin Kidd of the U.S. Army Command and General Staff College, Elizabeth Martin of the office of the Department of State's Coordinator for Reconstruction and Stabilization, and Colonel Jay Greer of the Department of State's Bureau of Political-Military Affairs.

Along the way, several experts gave generously of their time to discuss our project. We wish to thank Dennis Murphy and Bill Waddell of the Army War College, Dr. Ralph Doughty of the Command and General Staff College, Colonel Bruce Reider of the Center for Army Leadership, and Dr. Jake Kipp of the School of Advanced Military Studies.

Equally important to our study were the contributions of assignment officers at the U.S. Army Human Resources Command and the Army's Senior Leader Development Office. The surveys these officers completed were long, complicated, time-consuming, and tedious, a significant added burden in the midst of very busy jobs. The information they provided was, however, critical both for identifying the positions that could develop the knowledge, skills, and abilities in question and for estimating the number of officers over time that would be able to develop those qualities.

The team greatly appreciated our sponsor's active interest and support. Colonel Chris Robertson, Lieutenant Colonel Emory Phlegar, and Lieutenant Colonel Vince Lindenmeyer of the U.S. Army Human Resources Command's Officer Personnel Management Task Force made valuable intellectual contributions that helped shape the study, and worked tirelessly to ensure our access to interview subjects throughout the U.S. Army.

Finally, we owe a debt of gratitude to our colleagues here at RAND for their advice and expertise. Lynn Scott, Ray Conley, and Al Robbert made invaluable contributions to shaping our methodology. James Dobbins, Andy Hoehn, Michelle Parker, Patrick Gramuglia, and Frank Kingett all provided substantive input based on their extensive government experience in contingency operations and strategy development, while Melissa Bradley provided valuable insight about the design of our survey.

List of Symbols

DA	Department of the Army
DASD	Deputy Assistant Secretary of Defense
FM	Field Manual
GAO	Government Accountability Office
JDAL	Joint Duty Assignment List
JIIM	Joint, Interagency, Intergovernmental, and/or Multinational
JP	Joint Publication
JQS	Joint Qualification System
KATUSA	Korean Attached to the U.S. Army
KSA	Knowledge, Skill, and/or Ability
MNF-I	Multi-National Force–Iraq
NDAA	National Defense Authorization Act
NGO	Nongovernmental Organization
NIMS	National Incident Management System
O*NET	Occupational Information Network
OPMS	Officer Personnel Management System
PAM	Pamphlet
PRT	Provincial Reconstruction Team
S-3	Operations Officer

SAMS	School of Advanced Military Studies
USAID	U.S. Agency for International Development
USJFCOM	U.S. Joint Forces Command
XO	Executive Officer

CHAPTER ONE

Introduction

Background

The current security environment has significantly increased the salience of capabilities provided to Army operations by other services, other U.S. government agencies, and multinational partners. The Department of Defense has emphasized joint operations for decades, ever since the passage of the 1986 Goldwater-Nichols Act. Emphasis on interagency and multinational collaboration is of somewhat newer vintage. In each of these domains, however, actual operational practice has tended to advance faster than institutional support. During Operation Desert Storm, U.S. Central Command integrated joint operations. In Iraq and Afghanistan, brigade commanders command sizable elements from other services, respond to multinational commands, and integrate brigade combat team operations with the activities of other U.S. government agencies, of other national governments, and of nongovernmental organizations (NGOs). Further, as the events surrounding Hurricane Katrina demonstrated, Army forces must be equally prepared to conduct civil support operations to assist state and local authorities while respecting the complex thicket of law, policy, and politics surrounding the employment of federal military forces in a domestic context.

U.S. authorities have recognized these circumstances and moved to address them. The 2006 Quadrennial Defense Review called for the development of a corps of national security officers, and for increased emphasis on interagency coordination and collaboration in the professional development efforts of the various U.S. government departments and agencies (Rumsfeld, 2006). With the passage of the John Warner National Defense Authorization Act (NDAA) in 2006, Congress expanded the definition of "joint matters" to include cooperation with other government agencies, the armed forces of other nations, and even nongovernmental organizations (Public Law 109-364, 2006). Finally, the President promulgated Executive Order 13434 in 2007, the objective of which was to facilitate the development of the aforementioned corps of national security officers (Bush, 2007).

Several important questions nonetheless remain. First of all, while law and policy recognize the importance of competencies in the joint, interagency, intergovernmental, and multinational (JIIM) domains, guidance identifies these competencies in only a general way. More specific identification and description is one essential step in designing career development pathways that will provide an adequate inventory of officers prepared to operate in those domains. Second, though the 2007 NDAA and the resulting Joint Qualification System (JQS) aggregate the joint, interagency, and multinational domains together, it is not intuitively obvious that these domains resemble one another enough to justify undifferentiated career patterns. Third, to inform the development of career models, if not actual joint qualification, the Army must have some idea of what kinds of positions confer what kinds of experience.

To answer these and other questions, the U.S. Army Human Resources Command's Officer Personnel Management System (OPMS) Task Force engaged RAND Arroyo Center. To conduct the requested analysis, the research team addressed both supply and demand aspects of JIIM competency development. In conjunction with the OPMS Task Force, the team identified four research tasks:

- Identify and describe the knowledge, skills, and abilities in the JIIM domains required at successive stages throughout officer careers.
- Identify developmental experiences associated with the knowledge, skills, and abilities described above.
- Create a model for developing desired JIIM knowledge, skills, and abilities in senior leaders.

- Validate the model.

First, the team identified knowledge, skills, and abilities important for success in the JIIM domains by interviewing officers with recent operational experience, as well as individuals from other services, agencies, and nations with whom they worked. Using that information, the team queried officers with experience in officer assignments to identify the kinds of positions that strongly developed such knowledge, skills, and abilities. Next, we combined our analysis of the interview data with a literature review on the development of expertise[1] to construct alternative approaches to career development in the JIIM domains, and modeled the steady-state inventory of officers that the Army would be able to produce under the strategic conditions the Army expects to exist for the foreseeable future. Last, we assessed the feasibility of our proposed approaches to developing proficiency in the JIIM domains.

Defining Terms

For the most part, the definitions for each of the JIIM domains are intuitive and follow accepted joint definitions. Intuition can mislead, however, as it does in the case of the intergovernmental and multinational domains. Joint doctrine defines multinational as "Between two or more forces or agencies of two or more nations or coalition partners." Thus this term connotes interaction with official entities, and not simply dealings with foreigners. Joint doctrine does not define the term "intergovernmental," but defines an intergovernmental organization as an international organization with a charter or treaty, like the United Nations or the African Union. After seeking guidance from our sponsor, we decided to define "intergovernmental" in terms of the relationships between local, state, and federal entities in the context of support operations. The actual definitions we used are below:

---

[1] The research team's reliance on theories about the development of expertise does not imply any judgment about the degree of expertise required, which could range from novice to expert.

- **Joint:** activities involving two or more military services in pursuit of a common end.
- **Interagency:** activities involving two or more U.S. government agencies, including the Department of Defense, to achieve a common end.
- **Intergovernmental:** activities intended to coordinate efforts between the Department of Defense and local, regional, and state authorities.
- **Multinational:** activities that involve U.S. Department of Defense organizations with the military forces of other nations under the rubric of a contingent alliance or coalition.[2]

It is also necessary to define the terms "knowledge," "skills," and "abilities" in order to mitigate the risk that readers' intuition may mislead them. These definitions are taken from the U.S. Department of Labor's O*NET[3] questionnaires.

- Knowledge areas are sets of facts and principles needed to address problems and issues that are part of a job.
- A skill is the ability to perform a task well. It is usually developed over time through training or experience. A skill can be used to do work in many jobs or it can be used in learning.
- An ability is an enduring talent that can help a person do a job.

Distinguishing between skills and abilities poses difficulties for researchers and especially lay respondents. For our purposes, the key

---

[2] The definition of intergovernmental and multinational domains could be broadened very reasonably to include nongovernmental organizations, a conclusion we reached after having provided most of our respondents with the definition above.

[3] According to its website, "The O*NET Resource Center is home to the nation's primary source of occupational information. It serves as a central point of information about the O*NET program and is the main source for O*NET products, such as the O*NET database, Career Exploration Tools, User Guides, and development reports. It also provides the latest news on O*NET developments and product releases, and important links to related O*NET sites." National Center for O*NET Development, "O*NET OnLine." As of July 15, 2010: http://online.onetcenter.org/

distinction was that skills could be acquired and developed through education and practice, while abilities were more innate and enduring.

Research Approach

The research team developed its research methodology based on five hypotheses. The team felt that testing these hypotheses against available empirical data would answer the questions underlying our four research tasks.

- Performance in the JIIM domains improves with experience.
- Knowledge, skills, and abilities required in the JIIM domains differ at the strategic, operational, tactical, and institutional echelons in kind as well as in degree.
- Within each echelon, each JIIM domain requires qualitatively different knowledge, skills, and abilities.
- Each Army officer functional category requires qualitatively different knowledge, skills, and/or abilities in each JIIM domain.
- Experiences in one JIIM domain, or even in an "Army" billet, could contribute to development in another JIIM domain.

Identifying Knowledge, Skills, and Abilities Associated with the JIIM Domains

Overall, the research team approached our first research task by adapting applied cognitive task analysis. Specifically, the team performed what Militello and Hutton call a "knowledge audit," in which practitioners are asked to describe the knowledge, skills, and abilities that contribute to proficiency in their domain. Researchers then aggregate and analyze the results. We found this approach attractive because it promised reasonably accurate results but required little in the way of prior training (Militello and Hutton, 1998).

The hypotheses described above informed the development of our sampling plan and research protocol. The team felt that the best way to identify the knowledge, skills, and abilities Army officers actually needed in these domains was to ask people who had worked in such JIIM contexts. Moreover, it was important to interview not only Army officers, but also individuals from other U.S. government agencies, state officials, and officers from America's current multinational partners. To make sure we selected the right people, we tried to interview people who had worked in each domain, at each echelon, in each functional category. Graphically, our sampling plan looked something like a Rubik's Cube, as depicted in Figure 1.1.

In the end, we conducted forty-one separate interviews and twelve focus groups with over one hundred individuals. Most, of course, were active and reserve component Army officers who had succeeded in a JIIM context. We did speak to sixteen civilian officials, seven officers from other services, and five officers from other nations. This is a very small sample relative to the number of hypotheses and the absolute minimum required by the sampling plan. The small sample size invites further study and should induce caution in acting on this study's findings. The data provided by these interviews seemed to be generally

Figure 1.1
Sampling Plan

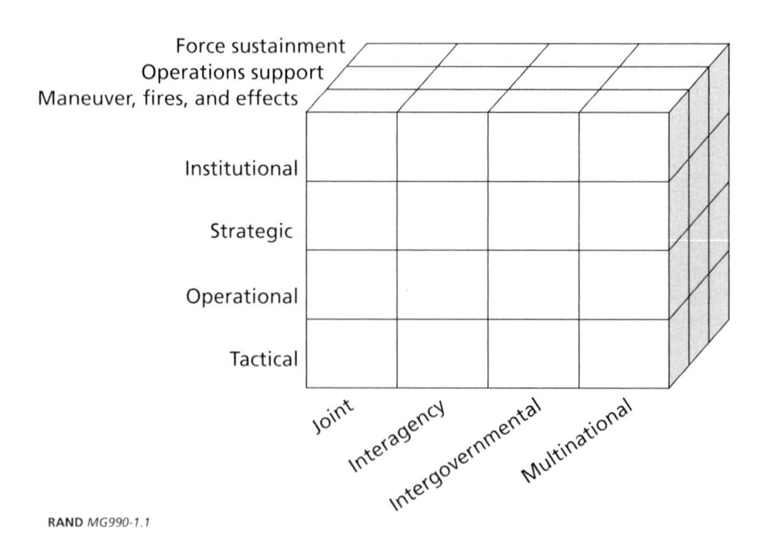

RAND *MG990-1.1*

consistent, however, allowing us to identify the range of possible knowledge, skills, and abilities required in the different JIIM domains.

In these interviews and focus groups, we followed a consistent protocol. We asked respondents to describe their duties, particularly the aspects of their jobs that were JIIM in nature, in order to allow us to decompose these tasks into supporting knowledge, skills, and abilities. Next, we asked them to describe the knowledge, skills, and abilities required to function effectively in these JIIM contexts. Finally, we asked them to describe the developmental experiences that had prepared them to perform their duties effectively and to identify the particular knowledge, skills, and abilities they had acquired in each experience. For a detailed description of the interview protocol, see Appendix B, "Interview and Focus Group Protocol."

After conducting and analyzing several interviews and focus groups, the team discovered that the questions about duties and tasks performed prompted respondents to identify and discuss the knowledge, skills, and abilities in question, obviating the necessity to decompose the tasks they described. The inquiries about developmental experiences had much the same result, in addition to identifying those JIIM opportunities. We also discovered that the same respondent's experience would cover multiple echelons and multiple domains. All this posed something of a challenge in attempting to code the resulting observations.

We then broke the data down into separate observations and coded them according to the JIIM domain, echelon, and functional category they addressed. An observation consisted of a statement identifying and describing a specific knowledge area, skill, or ability. Each observation was then aligned with a specific JIIM domain and echelon depending on its content and its context. For example, absent qualifying information, we would align an observation with the echelon and functional category held by the respondent. If the research team was interviewing an individual about his experience at the joint, strategic level as a logistician, his observations would be aligned with that functional category and echelon unless he or she attributed that observation to another context, e.g., "when I was a . . ." Overall, the team discovered over 900 separate observations that could be aligned with a JIIM domain, echelon, or functional category. We aggregated qualitatively similar observations into larger categories of JIIM knowledge, skills, or abilities. Observations like "you need to be able to listen" and "it helps to be able to get along with other people" might be integrated into "interpersonal skills." Ultimately, a set of 23 knowledge areas, 21 skills, and 6 abilities resulted from the aggregation process. We aligned a knowledge, skill, or ability with a particular domain, echelon, or functional category when at least half of all the observations about that knowledge, skill, or ability aligned with a particular domain, echelon, or functional category. Knowledge areas, skills, or abilities identified as important by a majority of our respondents but not systematically associated with any particular domain were classified as being important across all the JIIM domains.

Identifying Developmental Opportunities

Once the study team identified the knowledge, skills, and abilities associated with each JIIM domain, it was possible to identify the positions that could develop them. Our approach was to survey assignment officers about the extent to which the positions they managed developed the knowledge, skills, and abilities in question. Ideally, we would have surveyed the over 66,000 members of the Army officer corps on the knowledge, skills, and abilities required to perform their jobs effectively; such an approach would be more appropriate to a longer-running and more deeply detailed study. Assignment officers, however, are supposed to be experts on the positions they manage. Moreover, they are in continuous contact with the incumbents in those positions, and therefore might be expected to be aware of their current requirements.

We therefore asked assignment officers to rate the kinds of positions they managed against the knowledge, skills, and abilities we had identified. We did not, however, ask them to align these positions explicitly with any of the JIIM domains. Given the relatively high stakes attached to serving in a "joint" position, we felt that such explicit alignment would bias our results enough to make them unusable. Service in a "joint" billet is a prerequisite for promotion to flag rank, and several respondents asked how we would deal with the possibility that respondents might simply score all the positions they managed as "joint" in order to confer advantages on the officers they managed. More importantly, because we hypothesized that capabilities relevant to the JIIM domains might be developed in any job, we wanted our survey population to focus on the capabilities developed rather than domain alignment. Instead, we aligned positions with domains ourselves, based on what we had learned about how different knowledge, skills, and abilities aligned with different domains. We were careful to keep our analysis on the conservative side, identifying only those positions that strongly developed knowledge, skills, and abilities in the JIIM domains.

Developing Army Proficiency in the JIIM Domains

We began our research on developing JIIM proficiency by reviewing the literature on officer development and on expertise in general. The most salient conclusion we drew from the academic literature was that developing true expertise takes a long time: different scholars describe a "ten-year" or "10,000-hour" rule, both numbers indicating the amount of accumulated experience required to develop expertise in a given field of endeavor (Ericsson, 2006a, 2006b; Ericsson, Krampe, and Tesch-Römer, 1993; Lord and Maher, 1991; Bransford, 2000; Norman et al., 2006). More recent studies indicate that the variety of the experiences is at least as important as their accumulated length. Finally, we learned that formal education was a critical element in the development of expertise.

It is not clear, however, whether "expertise" is the right standard for development in the JIIM domains. First, the literature could not tell us whether the JIIM domains constituted independent fields of endeavor, like medical specialties, or refinements of officers' functional expertise, like competence in various software applications. It was therefore difficult to gauge how much effort was required to attain true "expertise." Second, it was far from apparent that true expertise was actually essential to functioning effectively in JIIM contexts; several of our respondents from other services, agencies, and nations asserted that it was more important that Army officers understood the Army well than for them to understand the JIIM aspects of a given assignment.

The research team chose expertise as a standard against which to evaluate the Army's ability to develop and maintain an inventory of JIIM-qualified officers because there is currently no other standard for identifying either the degree of expertise or the number of experts required. If it appeared that the Army could produce enough experts even when expertise required three or more assignments in a given JIIM domain, then the Army would certainly be able to produce enough officers with the desired degree of proficiency under less demanding criteria.

From prior RAND research, we derived general approaches toward developing an inventory of officers with the required JIIM proficiency. Thie et al. describe three basic approaches: a managing-skills approach, the intent of which is to distribute competencies in given areas as widely as possible throughout the force; a managing-competencies approach, the objective of which is to develop a cadre of experts in a given domain; and a leadership-succession approach, which is very similar to the managing-competencies approach, except that it focuses development on officers identified as potential senior leaders (Thie et al., 2005). The similarity between the managing-competencies approach and the leader-succession approach enabled us to restrict our modeling efforts to the first two approaches.

Finally, as aforementioned, we had included questions about the development of JIIM proficiency into our interview protocol. The most important thing our respondents told us was that the best preparation for service in a JIIM context was "something different." What they meant was having to work in an unfamiliar context, where one's success or failure essentially rested on the voluntary collaboration of individuals from different organizational or national cultures. They also indicated to us that the sequence of experiences did not particularly matter, though recent experience was to be preferred. Finally, many respondents with extensive JIIM experience indicated that one assignment would not suffice to develop adequate competence in a given JIIM domain. Respondents also indicated, however, that an officer's overall quality had more effect on performance in a JIIM context than the depth of his domain experience.

Because we could not definitively resolve either the question of just how much experience was required to develop the required degree of proficiency or the more important question of how much proficiency was actually required, the research team decided to model both a managing-skills approach and a managing-competencies approach. In the former case, the team sought to quantify how many colonels the Army could produce who had completed at least one tour in a JIIM assignment. In the latter case, we tried to quantify how many colonels the Army could produce who had accumulated three assignments in any one JIIM domain.

The study team adapted an officer inventory projection model developed earlier by RAND analysts. This model is a standard, steady-state inventory projection model that essentially performs linear optimization against a given set of criteria. RAND's model accounts for the complexities of modern Army officer management, such as different opportunities and outcomes in different branches and functional areas, the apportionment of branch-immaterial positions, and the migration of selected officers from basic branches to functional areas after about seven years of service. Its primary use is to compare the effects of different personnel policy options upon the composition and experience mix of the Army's future officer inventory. In this case we adapted its parameters to allow us to optimize either for breadth or depth of experience.

Organization of This Report

The report is organized into five chapters. Aside from this introductory chapter and Chapter Five, "Conclusions," each chapter follows a similar pattern. We begin by describing the particular research questions being addressed, then move on to our research methodology, and conclude with our principal findings. Three appendixes supplement these chapters and provide additional detail.

Chapter Two is our most important chapter. In it, we describe the research approach we used to identify and describe the knowledge, skills, and abilities associated with each JIIM domain and with the tactical, operational, strategic, and institutional echelons of military organization. In Chapter Three we identify the kinds of positions likely to provide experience in the various JIIM domains. Chapter Four discusses our extrapolations from the results described in Chapters Two and Three to estimate the Army's inventory of officers with JIIM experience or expertise. Finally, we restate and expand upon our principal results and conclusions in Chapter Five. We also describe possible avenues for further research.

A Note on Citations

This report relies heavily on interview and focus group data. We have listed these interviews, all of which occurred in 2008, in the bibliography. To avoid duplication and distraction, however, we provide a formal citation only when we do not mention the source in the text. Most of these interviews were with military officers and other officials, many of whom have since been promoted. The ranks and positions cited are the ones these individuals held at the time of the interview.

CHAPTER TWO

Identifying and Describing Knowledge, Skills, and Abilities Associated with the Joint, Interagency, Intergovernmental, and Multinational Domains

Introduction

Our single most important research task was to identify the knowledge, skills, and abilities (KSAs) that Army officers require and were acquiring in the JIIM domains. We wanted to focus on the knowledge, skills, and abilities that were unique to those domains, as opposed to the comprehensive set of competencies an officer serving in a JIIM billet might require.

While improving "jointness" has preoccupied the defense establishment for well over twenty years, there is relatively little in the way of literature that identifies and describes specific KSAs that are distinctly joint. Certainly, joint institutions for professional military education have spent considerable effort determining what soldiers, sailors, airmen, and Marines have to learn in order to be prepared for a joint assignment. These efforts are mostly deductive in orientation, deriving learning objectives from doctrine and the nature of joint capabilities. What has been missing is an inductive approach, rooted in organizational and industrial psychology, that sought to identify the required KSAs based on what officers actually did in their jobs. Recently, the Joint Staff J-7 commissioned a study to identify the competencies required in an accomplished joint warfighter. The study's focus was broad and included all competencies an officer required, making it difficult to discern those competencies that were narrowly joint in nature. Moreover, the comprehensive nature of a competency-based analysis complicates efforts to identify how elements of competency in one domain might support the attainment of competency in another. Decisions by Congress and the Department of Defense to broaden the definition of joint experience to include interagency, intergovernmental, and multinational affairs have further complicated the question, leaving the term "joint" to describe a wide range of activities that are actually significantly different. For that reason, we decided to do two things: take an inductive approach, and focus on KSAs that were specific to the JIIM domains.

Methodology

To identify the KSAs associated with the JIIM domains, we used an empirical approach based loosely on applied cognitive task analysis, a technique developed for the Navy Personnel Research and Development Center in 1998 to elicit task descriptions from expert practitioners in a cognitive domain. Specifically, we conducted what Militello and Hutton describe as a knowledge audit, in which information about domain-specific expertise is elicited by interviewing expert practitioners. In the interviews, the study team asks subjects to identify elements of expertise and then probes for concrete examples that illustrate the particular knowledge, skills, or abilities in question. In this study, we began by showing respondents a range of potential tasks in the various JIIM domains. We then probed to get respondents to identify the KSAs relevant to their performance of those tasks, and the experiences that developed those skills. During these interviews, the team took detailed notes; most of the interviews were digitally recorded as well.[1] The team then transcribed its notes to allow coding and subsequent analysis of the data (Militello and Hutton, 1998). As described earlier, our analyst then broke the interviews down into individual observa-

[1] In their initial foray into applied cognitive task analysis, Militello and Hutton found that good notes were as useful as exact transcriptions of the interviews.

tions and coded those observations according to the specific knowledge areas, skills, or abilities described, as well as the echelon and functional category in which they applied. We found this methodology attractive because it promised reasonably accurate results but required relatively little up-front training for the study team. This was an important consideration, as only one team member was in fact a credentialed behavioral scientist.

We structured our research plan to test five hypotheses about the nature and development of proficiency in the JIIM domains:

- **There are meaningful, qualitative differences between KSAs associated with the JIIM domains.** Put another way, the same set of KSAs will not serve equally well in all JIIM contexts.
- **These differences persist at the tactical, operational, and strategic levels of war, as well as the domain of military institutions that undergirds capabilities at all three levels.** The military recognizes three "levels of war": the tactical level, at which battles and engagements are fought; the operational level, in which battles, engagements, and other activities are arranged and sequenced in time and space in order to achieve strategic objectives in a theater of war or operations; and the strategic level, at which a nation or coalition establishes its objectives and employs its resources to attain those objectives (JP 3-0, *Joint Operations*). Although not doctrinally recognized as a "level of war," the military also devotes significant effort to the development and maintenance of military capabilities, an endeavor that calls on substantially different KSAs than the conduct of tactics, operations, or strategy; we chose to refer to that as the "institutional echelon." We hypothesized that even within a given JIIM domain, practitioners might require a different set of KSAs depending on the echelon at which they served.
- **There are meaningful, qualitative differences between KSAs associated with the JIIM domains associated with each Army officer functional category.** The Army organizes officer branches and functional areas into three broad categories: (1) maneuver, fires, and effects, which includes branches like infantry, armor,

and field artillery that tend to generate effects directly; (2) operations support, including officers who perform technical or cognitive functions that facilitate maneuver, fires, and effects, like signal, military intelligence, or operations research; and (3) force sustainment, including logisticians, finance officers, and others who manage human, materiel, and financial resources. At the outset, we thought it probable that even within a given JIIM domain, different functional areas might require different sets of JIIM knowledge, skills, and abilities.

- **There is a hierarchy of professional proficiency within each of the JIIM domains.** Simply put, there are measurable distinctions between experts and novices within each domain.
- **There are multiple developmental paths that can allow soldiers to attain the KSAs appropriate to each level of development, in each JIIM domain, in each functional category.** Based on our assumption that discrete knowledge, skills, and abilities support proficiency in each domain, we hypothesized that it might be possible to attain proficiency within a given domain by some path other than education and experience within that domain.

We held no strong assumptions about the truth or falsity of the above hypotheses. In fact, we strongly assumed that there might be significant overlap between the sets of KSAs associated with each JIIM domain. We did feel that data gathered to test these five hypotheses would provide sufficient bases to enable us to identify most of the KSAs in question and to discern how they were developed.

Data Elicitation

In the course of this study, we interviewed Army officers with experience at all levels in the various JIIM domains to identify those KSAs that were either unique to a particular JIIM domain, or uniquely important for success within that domain. For instance, many respondents asserted that "interpersonal skills" were extremely important in a JIIM context, even though "interpersonal skills" are important in many other fields of endeavor. More importantly, we also interviewed individuals from other services, U.S. government agencies, and nations to get their perspective on this question. Finally, we supplemented these interviews with focus groups at the Army War College, the U.S. Army's Command and General Staff School, NATO's Allied Command-Transformation, and the Department of State.

We selected respondents according to our research hypotheses. In order to find individuals who had served at the strategic, operational, and tactical echelons, we needed to associate organizations with those echelons. Figure 2.1 aligns the echelons of military command, federal departments and agencies, and state and local governments with the three levels of war. We interviewed individuals at all levels in the Department of Defense and from other executive branch agencies; we had less success identifying respondents with operational experience from state and local governments. It is important to note that all jobs at a given echelon may not share the same cognitive orientation. One need not be an accomplished strategist in order to contribute effectively at the strategic level.

As noted earlier, we also sought to investigate the KSAs associated with the institutional domain, in which we included activities whose primary purpose was the development and maintenance of military or civilian capabilities applicable to the conduct of operations in a JIIM context. In each JIIM domain, at each echelon, we also tried to get a perspective on each functional category. The list of individuals we interviewed and focus groups we conducted is included in the bibliography.

Our sampling plan could therefore be represented as a three-dimensional matrix as depicted in Figure 2.2. We tried to interview at least one incumbent and one individual from some external organization from each "cell" of the matrix. Often, our respondents were able to provide us with information relevant to other echelons and other domains, which helped shape our sampling plan and also contributed directly to our analysis.

Identifying experts proved to be a particular challenge, in that there are few formal, objective, and publicly available indices of expert performance in military operations. To the maximum extent possible, we tried to find people whose attainments made them recognized experts in their field. We also relied on those recognized experts' identification of other potential respondents. Still, as will be seen shortly, it

Figure 2.1
Comparison of U.S. Agency Organizational Structures

	Armed Forces of the United States	Executive Departments and Agencies	State and Local Government
Strategic	Secretary of Defense Chairman of the Joint Chiefs of Staff Joint Chiefs of Staff Combatant commander (1)	National headquarters Department secretaries Ambassador/Embassy (3)	Governor
Operational	Combatant commander Commander, Joint Task Force (CJTF) (2) Defense Coordinating Officer/Defense Coordinating Element	Ambassador/Embassy Liaisons (4) Federal Coordinating Officer or Principal Federal Official Regional Office	State Adjutant General State Coordinating Officer Office of Emergency Services Department/Agency
Tactical	CJTF Components Service Functional	Ambassador/Embassy Field Office U.S. Agency for International Development (USAID) Office of Foreign Disaster Assistance (OFDA) Disaster Assistance Response Team (DART) Liaison (5) Response Team U.S. Refugee Coordinator	National Guard County Commissioner Mayor/Manager County City (e.g., Police Department)

1. The combatant commander, within the context of unified action, may function at both the strategic and operational levels in coordinating the application of all instruments of national power with the actions of other military forces, U.S. government (USG) agencies, nongovernmental organizations (NGOs), regional organizations, intergovernmental organizations (IGOs), and corporations toward theater strategic objectives.
2. The CJTF, within the context of unified action, functions at both the operational and tactical levels in coordinating the application of all instruments of national power with the actions of other military forces, USG agencies, NGOs, regional organizations, IGOs, and corporations toward theater operational objectives.
3. The Ambassador and Embassy (which includes the country team) function at the strategic, operational, and tactical levels and may support joint operation planning conducted by a combatant commander or CJTF.
4. Liaisons at the operational level may include the Foreign Policy Advisor or Political Advisor assigned to the combatant commander by the Department of State, the Central Intelligence Agency liaison officer, or any other U.S. agency representative assigned to the Joint Interagency Coordinating Group or otherwise assigned to the combatant commander's staff.
5. USAID's OFDA provides its rapidly deployable DART in response to international disasters. A DART provides specialists, trained in a variety of disaster relief skills, to assist U.S. embassies and USAID missions with the management of USG response to disasters.

SOURCE: Joint Chiefs of Staff, JP 3-08 (2006).

RAND *MG990-2.1*

Figure 2.2
Sampling Plan

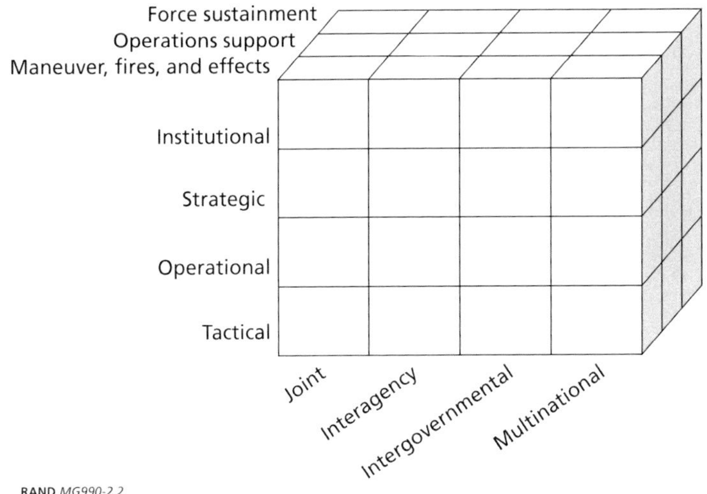

RAND *MG990-2.2*

proved difficult to find enough experts using these criteria, especially at lower ranks. We therefore included in our sample a great many individuals who simply had experience in these domains. Overall, we conducted interviews with 41 individuals and 12 focus groups with 2 to 8 participants each, for a total of 102 individuals from varying backgrounds.

Having identified our respondents, we conducted our interviews according to a standard protocol in order to elicit descriptions of the KSAs associated with each domain. In general, we asked subjects to describe:

- The titles and general duties of the job or jobs they held in a joint, interagency, intergovernmental, or multinational context.
- What was uniquely joint, interagency, intergovernmental, or multinational about those duties.
- The specific tasks they performed that were joint, interagency, intergovernmental, or multinational in nature.

- The KSAs associated with those joint, interagency, intergovernmental, or multinational tasks
- The experiences they felt developed the requisite KSAs.

When respondents were hesitant to name tasks or KSAs, we presented them with detailed potential task lists and KSA lists to give them an example of the range that could be addressed; however, most respondents were fairly fluent and did not require these aids. The reader will find these task lists included in a more detailed protocol in Appendix B. We also asked respondents to identify how long it took them to develop the relevant KSAs and to assess their relative importance. Naturally, not every interview followed the same lines. Almost invariably, respondents' relative emphasis on various KSAs provided an implicit assessment of their relative importance and helped us filter the important from the merely desirable. During the interviews and focus groups, team members took detailed notes, often supplemented by digital recordings, then transcribed those notes for subsequent analysis.

The data on the job requirements were collected at a relatively general level. Participants were instructed to think about the JIIM tasks they performed in the jobs for which they had been selected (i.e., jobs identified as having JIIM characteristics) in order to bring to mind actual activities performed, given that most of the participants were queried about positions formerly held. They were then asked what KSAs they found required for the job. The linkage to tasks attempted to invoke attributes at a level of detail and descriptiveness that would avoid some of the problems of vagueness and imprecision associated with the practice of competency modeling (e.g., see Sackett and Laczo's 2003 discussion).

We used O*NET KSAs[2] as the skeleton of our coding model and the basis of the instrument we used as a prompt when participants were having difficulty generating necessary attributes. As noted by Sackett and Laczo (2003), O*NET KSAs are at a level of generality that allows

---

[2] The Bureau of Labor Statistics in conjunction with several other research institutions have identified a standard taxonomy of knowledge, skills, and abilities that might be required in any given job. These standard lists, available at http://www.onetcenter.org/questionnaires.html, allow for comparison and differentiation between jobs.

for comparison across occupational areas. Realizing that these formalized knowledge, skills, and abilities were not designed to describe military jobs in a conflict environment, we then added KSAs implied by the JIIM tasks we had included on our task lists. The complete list of KSAs that we showed to respondents is included as Enclosure C to Appendix B, the interview protocol.

Although most of our respondents were Army officers or supervisors of Army officers, and we did distinguish by functional area, ultimately we were comparing across a breadth of occupational areas as well as jobs that change by their nature. For example, the course of an operation could require combat operations at some points and stability operations at others. Thus, we needed a certain amount of generality. We tried to strike a balance between generating a level of detail that would render relevant differences visible and accomplishing our goal of reaching a broad sample of high-level officers and government officials. Moreover, although we were comparing jobs, we were entering a relatively new domain and wished to allow our participants to dictate the content to the extent possible in order that we not skew the results by assuming *a priori* a model with more specificity than we actually had.

Our data should be regarded as just a sample of the potential variations in JIIM duties and positions. As this was primarily an exploratory study, we used qualitative techniques to get a rich level of detail on the positions that respondents were describing. It was impossible to interview incumbents from all of the possible permutations of echelon, functional category, and JIIM domain. Even if that had been possible, the nature of the jobs clearly changed from person to person and with the deployment location, if any, and the nature of the mission at a particular place and time. Thus, we used our purposive sampling strategy to illuminate potential variations across domain, echelon, and functional category rather than to permit broad statistical generalizations.

The majority of the information (i.e., 60 percent of KSA instances) was obtained via interview, reflecting the comparative richness and depth that can be elicited in that context. Certainly, Fern (1982) indicated that interview format elicitation generated more ideas than did the focus group format, so our finding that more information was generated during interviews is not surprising. The majority of intergovernmental KSAs were obtained in interview contexts, as were the majority of maneuver, fires, and effects KSAs. In general, however, the range and frequency of knowledge, skills and abilities identified did not differ significantly in the interview context than in focus groups. The only exception to this was the knowledge area of statutory, regulatory, and policy environment for homeland defense, which was more frequently elicited via interview. As that knowledge area was uniquely associated with the intergovernmental domain, as discussed below, and since the majority of the data elicited for this domain was from interviews, this finding also is not surprising.

Morgan (1996) reviewed the available empirical comparisons of the data elicited in individual interview and in focus group contexts and indicated that for some topic areas (e.g., sexual experiences) differences in the data itself (rather than just the quantity) could be anticipated. However, he indicated that differences might be expected to be a reflection of culture as much as method. Given that our topic area was relatively innocuous (presumably, *relatively* few cultural mores surround discussion of the KSAs required for Army officer jobs), and given that the primary method effect demonstrated was the quantity of data elicited via the two methods rather than systematic differences in the content elicited, we felt that the data were similar enough to be combined. The remainder of the analyses combined focus group and interview data. However, for the most part we avoided direct statistical tests because the exploratory nature of the study, broad topic area, subsequently broad sampling plan, and relatively small number of participants inherently limited the stability of our findings and made them less suited to broad generalizations across populations.

Data Analysis

We used the information collected in our interviews to examine three of our five hypotheses:

- There are meaningful, qualitative differences between knowledge, skills, and abilities associated with the JIIM domains.

- These differences persist at the tactical, operational, and strategic levels of war, as well as in the domain of military institutions that undergirds capabilities at all three levels.
- There are meaningful, qualitative differences between KSAs associated with the JIIM domains associated with each Army officer functional category.

We tested these hypotheses based on the frequency with which a particular knowledge, skill, or ability was associated with a given domain, echelon, or functional category. First, we associated each knowledge area, skill, or ability that emerged from our interviews with one of the JIIM domains. Next, we attempted to associate it with a particular echelon. Finally, we attempted to align the various knowledge areas, skills, and abilities with one of the three functional categories. Each of these analyses was independent from the others. That is, we did not classify a knowledge area, skill, or ability as being characteristic of a given JIIM domain and echelon and functional area. Given the low frequency of occurrence of the KSAs, we deemed it inappropriate to analyze more than one dimension at a time, as such fine parsing of the data might render our conclusions less stable.

In general, we aligned a particular knowledge area, skill, or ability with a given domain, echelon, or functional category only when that association was clear and unambiguous. For example, if it was unclear in which domain a necessary knowledge area, skill, or ability fell, we excluded it from our analysis. However, we did examine the frequency with which knowledge areas, skills, and abilities occurred by domain, by echelon, and by functional area. When a given knowledge, skill, or ability was mentioned more than twice overall and 50 percent of the time or more in conjunction with a given domain, we considered it to be associated with that domain. When all of the times a KSA was mentioned, it was in conjunction with a given domain, we considered it essential to that domain. Some skills, including general interpersonal skills and conflict resolution and negotiation skills, were mentioned with such frequency across JIIM domains that, although they were not primarily associated (in our sense of the word) with a given domain, we considered them essential for all. We performed similar analyses for knowledge areas, skills, and abilities by echelon and functional area. Finally, we checked our results by conducting an analysis of intra-rater reliability, or, in layman's terms, the degree to which our coder's assessment was consistent. Percentage agreement was acceptable given the complexity of the coding at 81 percent for the 50 aggregated categories.

Due to the aggregation, the same individual could have mentioned a given KSA multiple times; however, rather than restricting the number of times an individual could have mentioned a given KSA to once, we retained the multiplicity as an indication of the importance of the aggregated KSAs. Moreover, in some instances our participants were describing the usefulness of a given knowledge, skill, or ability in multiple jobs, and we wished to retain information about the largest possible number of positions held. However, in some cases we did examine the number of respondents who mentioned a KSA in our analysis, as a complementary indicator of importance.

Findings

The five most frequently mentioned knowledge areas, skills, and abilities overall were

- General interpersonal skills.
- Knowledge of other government agencies' capabilities, culture, and processes.
- Communication skills (both written and oral).
- Conflict resolution and negotiation skills.
- Knowledge of other services' capabilities, culture, and processes.

Perhaps our most significant finding, however, was that none of our respondents identified any particular knowledge, skill, or ability as being an absolutely critical characteristic, one for whose absence they could not compensate in a JIIM context. That is, lacking any one knowledge, skill, or ability was unlikely to cause any officer to fail in any given job, according to our respondents. Beyond that, respondents tended to accord more weight to general skills and abilities than to domain knowledge, according the greatest importance to "people skills," overall the most commonly noted KSA. In other words, skills in establishing relationships, in communication, and in negotiating with individuals from other organizations and influencing them largely sufficed for achieving success in a JIIM context. Moreover, the same sets of skills and abilities tended to be required in each JIIM domain, to varying degrees. That having been said, requirements for knowledge did tend to increase fairly dramatically in importance at senior levels. For instance, Colonel Fitz Lee, the Department of Defense representative at the Department of State's Coordinator for Reconstruction and Stabilization, asserted that understanding Department of Defense capabilities, organization, and processes was absolutely essential to doing his job. Moreover, the knowledge required tended to be the largest element of difference between the JIIM domains.

We confirmed our first research hypothesis: the JIIM domains are, in fact, distinct from one another. The four domains each required different combinations of knowledge, skills, and abilities, grouped as indicated in Table 2.1. The KSAs included in the table (labeled "K," "S," or "A") were those that stood out as associated with or essential to a given domain, as described above. Again, we associated a knowledge area, skill, or ability with a given domain when over half the relevant observations identified it as contributing to success in that domain; we classified it as essential if all observations identified it as contributing to success in that domain. This finding contrasts significantly with the current joint qualification system, which essentially treats all four of the JIIM domains as aspects of "jointness," on the basis that they are all components of unified action (U.S. Joint Knowledge Development and Distribution Capability, 2008). The existence of this difference suggests that officers serving in these different domains may require different developmental patterns.[3]

[3] Several of the skills and abilities listed in the multinational column of Table 2.1 would seem to apply more generally, e.g., "training management." Respondents, however, were identifying KSAs either unique to a particular JIIM domain or uniquely important. The relatively higher frequency of responses associated with the multinational context could stem from respondents' involvement in training indigenous security forces, which places a premium on basic, functional military competencies.

Table 2.1
Knowledge, Skills, and Abilities Associated with Each JIIM Domain

Joint	Interagency	Intergovernmental	Multinational
U.S. Army capabilities and doctrine (K)	Other government agencies' capabilities, culture, and processes (K)	Statutory, regulatory, and policy environment for homeland defense (K)	Area expertise (history, geography, culture) (K)
Joint capabilities and doctrine (K)	Cultural metaknowledge (K)	National Incident Management System (K)	Allied nations' capabilities, culture, and processes (K)
Joint organization and processes (K)	U.S. government strategy and policy (K)		Partner nations' capabilities, culture, and processes (K)
Strategic issues (K)	U.S. government law, policy, and processes for allocating resources (K)		International and nongovernmental organizations' capabilities, culture, and processes (K)
Other services' capabilities, culture, and processes (K)			Stability and counterinsurgency theory and doctrine (beyond official U.S. doctrine) (K)
U.S. Army structure, processes, and culture (K)			NATO capabilities, culture, and processes (K)
Joint planning processes and system (S)			Active/self-initiated learning (S)
Management of financial resources (S)			Change management and project management (S)
Originality (A)			Employing U.S. Army capabilities (S)
			Instructing (S)
			Judgment and decisionmaking (S)
			Management of personnel resources (S)
			Training management (S)
			Comfort with ambiguity/adaptability (A)
			Flexibility (A)
			Conscientiousness/integrity/decisiveness (A)
			Deductive/inductive reasoning (A)

We also confirmed our second research hypothesis: different sets of JIIM knowledge, skills, and abilities were required at the echelons habitually associated with the three levels of war and in military institutions. For instance, an officer serving as the operations officer in a joint task force needs to understand joint capabilities and doctrine, while it is more important for an action officer on the Joint Staff to know joint organization and processes. The association of distinct KSAs with different echelons is shown in Table 2.2 below.

A brief look at Table 2.2 indicates that at least some of these associations are counterintuitive, at best. Probably the most valid conclusion to be drawn is that JIIM knowledge, skills, and abilities do in fact differ by echelon, though perhaps not in the ways identified in Table 2.2. Our respondents' subjective assessments supported our quantitative analysis. Several respondents with experience at several

Table 2.2
JIIM Knowledge, Skills, and Abilities Associated with Echelon

Strategic	International and nongovernmental organizations' capabilities, culture, and processes U.S. government strategy and policy Strategic issues Other services' structure, processes, and culture U.S. Army structure, processes, and culture Social perceptiveness Comfort with ambiguity/adaptability Flexibility
Operational	National Incident Management System NATO capabilities, culture, and processes
Tactical	Area expertise Stability operations and counterinsurgency theory and doctrine (beyond U.S. doctrine) Cultural metaknowledge Employing Army capabilities Coordinating with personnel from other nations Employing joint capabilities Foreign language skills Stress management Self-awareness Training management

echelons agreed with the assertion that there were significant differences between the JIIM knowledge, skills, and abilities required at different echelons, some emphatically; none disagreed (Army War College focus group 2, Hoffman, Zabel, Grimes, Anderson, MacFarland). Their first-hand experience of the differing requirements posed at different echelons carries special weight. Moreover, while combining the results of Table 2.1 with Table 2.2 to form a matrix aligning knowledge, skills, and abilities by domain and echelon might recommend itself as intuitively obvious, unfortunately this would not be a valid approach. This is because such fine parsing would require reliance on comments made by very few of our participants in many instances, so that any firm conclusions would be an over-extrapolation.

We were unable to differentiate the KSAs required in the JIIM domains by Army officers' functional category. That is, our analysis did not uncover any difference between the knowledge, skills, or abilities required by maneuver, fires, and effects officers, operations support officers, and force sustainment officers working in the same domain, at the same echelon.[4] At least, we were unable to discern any differences that made sense. For instance, "instructing" appeared to be uniquely aligned with the force sustainment functional category. This finding may have more to do with our method than the actual truth of the matter. Our interviews were too short to penetrate to the details of an individual's job. For example, Brigadier General Stephen Anderson, the former Director of Resources and Sustainment for Multi-National Force–Iraq, noted that he had to understand other services' sustainment requirements, while Colonel Sean MacFarland said that his brigade staff had to work through differences in Army and Marine Corps airspace control measures. Because there were not enough of such observations, all we could conclude with any confidence was that officers needed to understand joint capabilities pertinent to their functional category. Such differences are amply accommodated by current assignment and education patterns.

---

[4] Statistical analysis did suggest alignment of some JIIM knowledge, skills, and abilities with functional categories, but those results were highly counterintuitive.

That is not to say that functional proficiency was unimportant. Indeed, the very opposite was true. Almost all of our respondents agreed with one focus group at the Center for Army Analysis who strongly maintained that success in JIIM contexts depended first on their general qualities as Army officers and next on their level of functional proficiency as analysts. KSAs unique to the various JIIM domains ranked third. The planners with whom we spoke were unanimous in asserting that well-developed Army planning skills were extremely useful in integrating JIIM activities. Several asserted that these skills were not only very useful but in fact sufficient for success in JIIM contexts. Colonel Sean MacFarland emphasized that his contributions in the JIIM domains fell well within the bounds of command appropriate to his level. From an outside perspective, Brigadier General Hussain Al-Sondani, formerly the assistant to the Iraqi Vice Chief of Staff, noted that most of the friction in his interaction with American officials stemmed from their highly variable understanding of force management, not from language or cultural issues. This conclusion accords with that of a Caliber Associates study for the U.S. Joint Staff J-7 on joint leader competencies, which found that the foundation of an effective joint officer was an effective service officer (Morath et al., 2007).

Lieutenant Colonel Tom Goss made what may have been the most important observation when he noted that the most important element of success in the JIIM domains was simply to realize that he was in a different environment, with different dynamics, which would require different behaviors. This observation applied to his experience on the U.S. Northern Command Staff and on the NATO international military staff. A signal officer in one of our Army War College focus groups made a similar observation. He had commanded a combined NATO communications organization, staffed with personnel from several countries. He felt that his effectiveness depended on recognizing that there were several perspectives on how to approach a given challenge, all of which might be valid (Army War College focus group 1).

The following sections describe first, the KSAs generally required in all the JIIM domains, and then those KSAs unique to each JIIM domain.

Knowledge, Skills, and Abilities Required Across the JIIM Domains

Over half of our respondents identified people skills, emphasizing the development and maintenance of relationships as the single most important KSA in the JIIM domains; about half of these respondents specified multiple different aspects of interpersonal skills, emphasizing its importance to them. According to Lieutenant General Frank Kearney, currently Deputy Commander of U.S. Special Operations Command, "JIIM is easy—it's about trust and relationships. It has nothing to do with knowledge." While neither we nor all of our respondents agree with General Kearney that success in a JIIM context has "nothing to do with knowledge" (Kearney, 2008), most respondents attached significantly higher importance to people skills than to domain specific knowledge.

Clearly, the term itself is somewhat imprecise, but in the context of our interviews and focus groups, our respondents seemed to define people skills as those that allow an individual to foster positive interaction with his or her counterparts and co-workers. Building and maintaining relationships appears to be a major aspect of people skills. According to the Army's Field Manual (FM) 6-22, *Army Leadership* (2006), relationship building is a technique in which practitioners build positive rapport and a relationship of mutual trust, making counterparts more willing to support requests. In fact, our respondents' description of people skills resembled FM 6-22's description of "influence techniques," especially those that did not rely on coercion or compulsion.[5] We should note that these influence techniques rest on a firm academic foundation, as a brief scan of FM 6-22's nonmilitary bibliography will reveal.

Negotiation ranked a close second in number of affirmations. Approximately 40 of our respondents noted that JIIM circumstances require cooperation and collaboration between many different organizations, with different national and organizational cultures and different

---

[5] While many respondents asserted that people skills are innate, they can in fact be taught and improved. Such training has been incorporated into the Transition Team Training Course at Fort Riley, Kansas. Lieutenant Colonel Chris Wilbeck, "Transition Team Training," PowerPoint briefing at the Joint Center for International Security Force Assistance, Department of Defense Advisor Training Working Group, October 27 and 28, 2008.

objectives. In the words of one student at the U.S. Army Command and General Staff School, "Everything's a negotiation" (U.S. Army Command and General Staff School Focus Group 1, 2008). Like there is for people skills, there is extensive academic research on negotiation theory and practice, much of which can be found at the Harvard Negotiation Project's web site.[6] As experienced Army officers can attest, negotiation is a skill with considerable application inside the Army as well.

Almost as many respondents thought critical thinking and analytical skills were important. This emphasis on critical thinking is, of course, a staple of all recent studies on military professional development and knowledge work in general. In spite of the utter predictability of its inclusion, critical thinking is genuinely important in JIIM contexts because many of the problems officers encounter are both knotty and unfamiliar. There were about half as many affirmations (37) of the importance of critical thinking as there were for people skills. In other words, our respondents seemed to think that people skills were substantially more important to success in JIIM contexts than critical thinking.

Respondents who worked at the strategic echelon were generally more likely to attach great importance to critical thinking and analytical skills. According to O*NET, critical thinking consists of "Considering the relative costs and benefits of potential actions to choose the most appropriate one. Analysis of complex problems to determine appropriate solutions" (O*NET, 2003). Tim Hoffman, the Director for Security Cooperation in the Office of the Under Secretary of Defense for Policy, went so far as to say that an officer with high intelligence and a rigorous graduate school education would probably be more useful for analyzing strategic issues in the course of working staff actions than a more senior officer who was simply a War College graduate.

From the study team's perspective, Hoffman's observation highlights an important nuance related to work at different echelons. The tasks performed at the strategic echelon do not consist solely or even principally of developing strategy. Officers working in the strategic domain typically perform tasks that enable others to develop strategy.

---

[6] http://www.pon.harvard.edu/research/projects/hnp.php3, accessed November 18, 2008.

Staff officers analyze issues, write papers, and present briefings. Critical thinking skills are obviously useful in these tasks, and are important in formulating strategic courses of action as well. They are hardly sufficient for the development of proficiency at the strategic level, however.

Skill in written and oral communication was closely linked with critical thinking skills. Respondents in every domain, in every echelon, felt that it was important to understand their audience and to convey facts, concepts, and plans in a manner that their audience would understand. Even more fundamentally, effective communication skills require practitioners to accumulate and present facts, assumptions, and conclusions in an orderly, logical manner. Communication skills are, of course, important for all officers. They assume particular importance in JIIM contexts, in which participants probably share neither the same lexicon nor the same worldview. Similar numbers of respondents indicated that communication skills were important as indicated that critical thinking skills were important; and these skill sets received approximately the same number of affirmations as well.

We should not close without noting that many respondents (about 35) felt that functional proficiency formed the foundation of their effectiveness in the JIIM domains. Respondents' strengths in the KSAs we describe enabled them to successfully apply their proficiency as infantry officers, engineers, analysts, and so forth in JIIM contexts. Functional proficiency, combined with people skills, allowed officers to successfully overcome shortcomings in other KSAs associated with the JIIM domains. The converse was not true, however. No degree of specific JIIM competency could overcome incapacity in an officer's functional domain.

Joint

Predictably, what distinguished the joint domain from the other JIIM domains was the knowledge required. A thorough understanding of joint organization and processes, combined with an equal degree of understanding of other services' capabilities, culture, and doctrine, greatly facilitated success in the joint domain. As we have noted, respondents tended to feel that such understanding was less important than people skills which enabled effective collaboration. That does not mean that knowledge was unimportant, however. For example, one officer recounted the story of his attempt to employ a B-52 strike to clear an enemy minefield during the first Gulf War. While the strike went in as planned, it did little to actually clear the minefield, something the officer later learned was entirely predictable (Quantock, 2008).

The nature and degree of understanding required varied by echelon, however. Officers serving at the operational and tactical levels required a fairly detailed, intuitive understanding of joint and service capabilities, while those at the strategic level, like Tim Hoffman, needed to know just enough to understand when other service representatives were making unrealistic proposals.

While understanding other services' capabilities was useful, understanding the U.S. Army's capabilities, culture, and processes was essential. Former Deputy Assistant Secretary of Defense (DASD) for Strategy Andy Hoehn wanted his subordinates to be able to articulate their service's position on any given issue, and to navigate the service bureaucracy. Army officers could compensate for shortfalls in knowledge of other services' capabilities through collaboration. In collaborative joint processes, officers from each service bring that knowledge to the table. Several respondents also noted that it was important for U.S. Army officers to understand the distinctive aspects of U.S. Army culture in order to identify potential friction points. To enable successful joint planning and execution, Army officers have to bring at least a working understanding of the availability and utility of the full range of Army capabilities.

Not surprisingly, respondents in interviews and focus groups at the strategic echelon tended to think it was important to comprehend the various strategic issues at play at any given time. By strategic issues, we mean the major strategic problems confronting the United States. This observation may be a function more of the echelon at which our respondents worked (generally in the Office of the Secretary of Defense) than of their association with the joint domain. In other words, while the office was joint, the issues were not necessarily so, nor even indisputably military in nature.

Competence in joint planning processes and supporting systems facilitated effective collaboration. There are two aspects to joint planning processes. One is the general analytical approach to understanding military problems. JP 5-0, *Joint Operations Planning* (Joint Chiefs of Staff, 2006), describes this approach, which resembles the Army's military decisionmaking process very closely. Army planners with joint and interagency planning experience told us that the military decisionmaking process described in FM 5-0, *Army Planning and Orders Production*, provided an excellent basis on which to conduct joint and interagency planning. Several Army respondents told us Army officers were perceived to have a comparative advantage as joint planners, in fact. Beyond that conceptual approach, however, there are the actual processes by which plans and ideas are translated into execution. The voluminous Chairman, Joint Chiefs of Staff series of manuals on the *Joint Operation Planning and Execution System (JOPES)* (CJCSM 3122) describes these processes. Put another way, the general conceptual approach described in JP 5-0 helps planners discern what they need to do and what capabilities they need to do it, while the systems and processes described in CJCSM 3122 are required to actually obtain and employ the required capabilities.

Respondents also cited the need for originality at higher levels. The Department of Labor defines originality as "The ability to come up with unusual or clever ideas about a given topic or situation, or to develop creative ways to solve a problem" (O*NET, 2003) Former DASD Andy Hoehn felt that officers working at the strategic level in the Office of the Secretary of Defense had to be able to develop "original content." What he meant was that action officers had to be able to conceive of novel applications of national power to particular strategic challenges. Colonel "Mo" Morrison, a military intelligence officer with extensive joint operational experience, said that successful joint intelligence officers needed to be able to "think the extraordinary." This meant developing and applying novel analytical frameworks to unprecedented situations.

Interagency

The interagency domain required KSAs similar to those required in the joint domain. Instead of understanding other services' culture, capabilities, and processes, officers working in this context had to understand the culture, capabilities, and processes of other government agencies. Respondents thought that a working knowledge of how the U.S. government allocates resources and responsibility was important, since deciding which agency pays for which activities is often the first order of business.

Interestingly enough, some respondents felt that a general appreciation for unfamiliar cultures and a thorough understanding of the culture, geography, and politics of the area of operations were important to success in the interagency domain. While such knowledge pertained neither to the workings of other U.S. government agencies nor to interagency management processes, respondents felt that this contextual understanding was essential to integrating civilian and military efforts.

Eighty-eight percent of specific observations pertaining to the interagency domain identified understanding other agencies' culture, capabilities, and processes as important. Colonel Don McGraw, who was the Director of Operations (CJ-3) for Combined Forces Command–Afghanistan in 2005, thought such understanding was vital. In order to integrate military operations with those of the multifarious U.S. government agencies operating in Afghanistan, he had to know how those agencies organized their efforts abroad and what they were doing, but was forced to acquire this knowledge on the job. Michelle Parker, who served as the U.S. Agency for International Development (USAID) representative on the Jalalabad Provincial Reconstruction Team (PRT) in Afghanistan, thought it imperative for military officers to understand other U.S. government agencies' charters and capabilities. Parker thought that friction over roles and authorities frequently hindered counterinsurgency and development efforts. This need for understanding other agencies extended to quite low echelons. For example, officers with experience on brigade combat team staffs in Iraq also cited their need to understand what other agencies did and how they might contribute to ongoing security and reconstruction efforts (U.S. Army Command and General Staff School Focus Group 1, 2008).

As noted, understanding how the U.S. government resources its activities abroad was important for success in unified action. This is particularly true since U.S. government processes and authorities for resourcing integrated operations are somewhat immature. The U.S. government funds operations in Afghanistan and Iraq using Title 22 (Foreign Assistance) monies, Title 10 (Defense) monies, monies appropriated for counternarcotics operations, and monies appropriated specifically for operations in those places, just to name a few sources. Each has certain restrictions placed on it. For example, foreign assistance funds cannot generally be used for the purchase of arms and equipment, at least without a waiver. Mac McLauchlin, at that time the Senior Advisor to the U.S. Ambassador to Afghanistan, got his job when he was first sent to Afghanistan in the course of an investigation into allegations of financial malfeasance arising from the use of Foreign Assistance funds to equip the Afghan Army. Very early in the investigation, both Ambassador Zalmay Khalilzad and the U.S. commander, Lieutenant General David Barno, recognized that there was no one to oversee all various U.S. government funding streams flowing into Afghanistan and so retained McLauchlin in that capacity. Brigadier General Stephen Anderson, formerly the Director for Resources and Sustainment at Multi-National Force–Iraq (MNF-I), made a similar observation. Michelle Parker provided another example when she described the integration of military commanders' emergency response program (CERP) funds with USAID resources as an example. PRT commanders would use CERP funds to initiate projects, because they could obligate those funds quickly, while USAID would go through normal budget processes to fund those projects through to completion (Parker, Ruf).

Several key respondents asserted that understanding the culture, geography, and politics of the area of operations was extremely important to success in the interagency arena. Ron Neumann, formerly the U.S. Ambassador to Afghanistan, was one such individual. Neumann felt that neither military nor civilian officials could contribute usefully to planning and conducting operations unless they thoroughly understood the operational context. Neumann felt that repetitive tours in a given area of operations were probably required to develop this level of proficiency. Similarly, Lieutenant Colonel Jim Ruf, who had been the military commander of the Jalalabad PRT, felt that doing a good job required in-depth study of the area of operations. According to Ruf, that meant understanding not just the country but also the locality in which one would operate.

On the other hand, Jim Dobbins, who was the Bush administration's first post-9/11 envoy for Afghanistan and who had coordinated U.S. efforts in the Balkans in the 1990s, felt that such area expertise was somewhat important, but that functional expertise on how to best employ instruments of influence was even more essential and often in shorter supply. In Dobbins's view, the challenges U.S. authorities faced were infinitely varied and complex, but the U.S. government possessed only a limited range of tools with which to address those challenges, to include economic assistance, diplomatic persuasion, and military compulsion. In Dobbins's view, there were usually experts available who could explain why a society was in conflict, but a dearth of those who knew how to end it. What U.S. officials needed most was to know how to best apply the limited tools at their disposal to effect the desired changes in other societies. An accurate diagnosis (regional expertise) was important in any such endeavor but skilled surgery (operational expertise) was equally so, and this was, in his judgment, an even rarer quality.

Intergovernmental

We were not able to elicit much input about the KSAs required in the intergovernmental domain, but the input we did receive was of very high quality. Beyond fairly obvious requirements, like the need to understand the statutory, regulatory, and policy environment for homeland defense, respondents stressed the need to forge and maintain personal relationships. Lieutenant General Clyde Vaughn, the Director, Army National Guard, emphasized the fact that each of the 54 states and territories has its own unique constitutional, political, and cultural context. While not as alien as operational environments abroad, the very familiarity of domestic environments might lull Army officers into ignoring important distinctions. Captain Rob Billings, Plans Officer for the State of Louisiana, provided a complementary view by noting that these distinctive features can be easily integrated into standard Army planning methods, if the planner is sensitive to their existence.

Lieutenant General Vaughn strongly emphasized the importance of personal relationships and of understanding the distinctive constitutional arrangements of the 54 U.S. states and territories in synchronizing and integrating activities in an intergovernmental context. Civil support operations have intensely political overtones, especially in the wake of natural disasters, and military officials need to be careful not to trespass on either the authority or the prestige of state and local officials. Supporting those officials effectively requires an understanding of both their official responsibilities and their personal strengths and limitations.

Respondents felt that a thorough understanding of the statutory, regulatory, and policy environment for homeland defense was very important in the intergovernmental domain. There are important constraints and limitations on what military forces can and cannot do in a domestic context. This environment includes U.S. Code, including Title 10, Title 32 (Reserve Components), the National Incident Response Plan, and other federal and local policies. This applies not only in support operations, but in the more routine activities of the U.S. Army Corps of Engineers (USACE). Major General David Fastabend, formerly the commander of USACE's Northwestern Division, cited the example of an engineer officer with the responsibility to manage a river with eight different fundamental purposes (irrigation, navigation, flood control, etc.), each purpose having a different collection of stakeholders. In such a case, the officer has to be able to reconcile the interests of these different stakeholders within the law, which means he or she must know the law.

Several respondents cited a need for practitioners to be able to function in the context of the National Incident Management System (NIMS), including the Incident Command System (Kingett; Dolan). The Incident Command System determines which agency is in charge, under what circumstances. According to Colonel (ret.) Bill Dolan, who led U.S. Joint Forces Command's (JFCOM's) study of the military response to Hurricane Katrina, that responsibility shifts, sometimes unexpectedly. Dolan felt that officers with homeland defense responsibilities needed to understand that system.

Captain Rob Billings, then the Plans Officer for the State of Louisiana, offered an interesting perspective on the intergovernmental context. Billings was responsible for leading the joint planning group responsible for coordinating military support to state and local agencies in the event of various contingencies. Like many of our Army respondents, Billings thought that the aforementioned knowledge of states' administrative organizations and other law and policy surrounding the civil support mission were merely planning factors that a skilled planner could integrate without extensive foreknowledge, though nonetheless important. He felt that a thorough understanding of Army capabilities, combined with skilled application of the military decisionmaking process, would allow successful management of both the legal and regulatory environment and the integration of other agencies' capabilities.

Multinational

Interestingly, most of the observations we recorded pertained to the multinational domain. According to Lieutenant Colonel Tom Goss, an Army strategist on the NATO international military staff, the key to success in the multinational environment was simply being aware that he was in a different environment and being willing to adapt to its dynamics. Obviously, respondents found it useful to have enough knowledge of allies' and partners'[7] capabilities and culture to be able to envision their ability to contribute to a particular operation or to anticipate their reaction to a given initiative. Similarly, several observed a need for skills in cross-cultural communication. And, while understanding NATO doctrine and processes may seem an equally obvious asset, it still bears explicit mention because NATO doctrine and processes have an impact well beyond operations conducted under NATO's aegis. Less obviously, respondents asserted that a broad understanding of theory and doctrine on counterinsurgency and stability operations facilitated interaction with the bewildering array of national, international, and nongovernmental entities with which they had to deal.

Oddly, respondents found that activities in a multinational context required them to be able to coordinate with personnel from other U.S. organizations at a higher rate than did activities in the joint, inter-

---

[7] Throughout the study, we distinguished between allies, with whom the United States shares a formally defined and long-standing relationship, and partners, with which the U.S. relationship may be of more recent vintage or more temporary in nature.

agency, or intergovernmental domains. The multinational context also seemed to call more heavily on generic skills and abilities than did the other JIIM domains. "Active learning" was very important, as were "instructing," "adaptability/flexibility," "deductive reasoning," and "management of personnel resources." The nature of current operations, which are heavily multinational at virtually every level, probably skews our findings. It also seems probable that the ambiguity and variability of the multinational context calls most heavily on general skills for understanding and adapting to unfamiliar contexts.

Several respondents agreed with Colonel Goss's observation that officers needed to start with the awareness that they were in a different environment and being willing to respond to it. Brigadier General Mike Ryan, the Director of Operations for the Allied Rapid Reaction Corps, started his interview by emphasizing the need for U.S. Army officers to realize that there were different cultures besides that of the U.S. Army, and being willing to work with people from those different cultures on their own terms. According to Goss, Ryan, and others, that awareness required listening very carefully to individuals from other organizations, and being willing to evaluate their input with an open mind. Such willingness did not require being "mushy." Brigadier Phil Jones, a British Army officer who had served as the Director for Strategic Policy and Planning (J-5) of Combined Joint Task Force-180 in Afghanistan, recalled one U.S. subordinate who very much fulfilled the stereotype of the hard-edged, hard-driving U.S. Army officer. This subordinate was nonetheless able to work well with officers from other nations because he was willing to take their input seriously.

That is not to say that a willingness to adapt is all that is required for success in the multinational domain. Lieutenant Colonel Dave Toczek, formerly of the J-35 (Plans) for the International Security Assistance Force in Afghanistan, said that understanding NATO doctrine was essential, because NATO allies will usually take action only in accordance with NATO doctrine. If NATO doctrine does not prescribe a certain activity, NATO members will be reluctant, if not unwilling, to perform that activity (Toczek, Goss, Ryan). NATO doctrine has importance beyond the NATO context, however, since its many member nations use its doctrine and processes in other NATO activities. For instance, the French Force Headquarters at Creil, the French Army's standing joint task force headquarters, uses NATO doctrine and staff procedures (Neveux, 2004; France, Ministry of Defense, 2009). Nonetheless, understanding NATO doctrine is not necessarily a prerequisite for assignment to a NATO position. According to allied officers at Allied Command-Transformation, officers can usually acquire such knowledge in the course of the first six months of an assignment.

Many respondents cited the need for a broad understanding of counterinsurgency and stability operations theory and doctrine. The literature on the subject is quite extensive, and there is broad agreement on core principles. Some, like Colonel Sean MacFarland, who commanded the 1st Brigade, 1st Armored Division in 2006, needed a broad understanding because formal U.S. doctrine was still maturing while his brigade was pacifying Ramadi. In other cases, however, such understanding enables effective collaboration with other partners who may or may not accept the authority of U.S. doctrine. Brigadier Jones observed that some of the most capable counterinsurgents, with the most extensive experience, were United Nations political officers. Ambassador Neumann felt that officers who were well-grounded in stability operations and counterinsurgency theory doctrine were usually well-prepared for the interagency environment as well, perhaps because of the emphasis these operations place on interagency cooperation.

The multinational context seemed to draw heavily upon generic skills and abilities, probably because of the heavily contingent nature of multinational operations. Perhaps the one most in need of clarification is "management of personnel resources." Many of our more senior respondents handled the multinational aspects of their jobs by assigning talented subordinates to manage them. Major General David Fastabend, describing his experiences as the Director of Operations for MNF-I, noted that he assigned his most talented subordinates to serve as liaison officers to Iraqi officials. According to Fastabend, managing these relationships was absolutely essential. On the other hand, identifying who would do a good job was difficult. First, he would screen people based on functional proficiency and select someone from that pool who possessed some ability to navigate Iraqi culture. Those officers would then serve in a probationary capacity. If the relationship did not work out, for whatever reason, Fastabend would quickly replace the officer. Colonel Jay Christensen did much the same as the Director for Sustainment (CJ-4) for Multi-National Corps–Iraq.

Finally, understanding and being able to navigate U.S. rules governing the transfer of classified materiel was a sensitive issue with many allied officers, though this area of knowledge was not necessarily a distinguishing trait of the multinational domain. Successful multinational operations require sharing information, but classification rules are designed to protect information, sources, and methods by restricting access. If U.S. officials lack a firm grasp of classification rules and foreign disclosure procedures, it can result in overclassification and shut down the necessary flow of information. Brigadier Jones noted that foreign officers working in U.S. headquarters were often shocked at the way the "door was slammed in their face" because of classification issues. Officers who had worked as operations research analysts on bilateral acquisition projects also observed that navigating classification rules was a significant challenge (Center for Army Analysis focus group, 2004). This is not to argue that classification rules are too restrictive, but that U.S. officers working in a multinational context need to understand them well enough to share information as well as protect it.

Compensating Competencies

As the study progressed, we began to note a striking similarity between our findings and that of another RAND study, *Compensating for Incomplete Domain Knowledge* (Scott et al., 2007). In this study for Project AIR FORCE, a RAND research team investigated the question of how Air Force general officers were able to manage large and complex enterprises with which they had had little prior experience. For example, how could an officer who had risen as a fighter pilot cope with the challenge of managing the Air Force Materiel Command? The RAND team found that such officers applied a number of competencies, summarized in Table 2.3.

These competencies fell into four broad categories: enterprise knowledge, integration skills, problem-solving skills, and people skills. "Enterprise knowledge" refers to understanding of the overall goals being pursued, and the role one's organization played in support of those goals. "Integration skills" refer to the ability to identify the right sources of information and analysis to bring to bear on a particular problem, and the ability to integrate relevant outputs into a solution. "Problem-solving skills" are those general skills which can be applied to any given problem, including defining the problem, establishing facts, identifying relevant analytical frameworks, and so forth. "People skills" means the general ability to foster effective collaboration on a particular issue.

As noted, there are several parallels with the KSAs that our respondents identified in the JIIM domains. "Enterprise knowledge" corresponds to observations on the importance of understanding stability operations and counterinsurgency theory and doctrine, particularly as it describes the roles and functions of civilian and military organizations and efforts. "Integration skills" and "problem-solving skills" would resonate with the significant and determined minority who

Table 2.3
Compensating Competencies

Enterprise knowledge • Organizations • Processes • People • Weapon systems	**Problem-solving skills** • Problem definition • Solution development • Knowing when to decide • Delegating to the right talent
Integration skills • Capabilities • Systems • Operations • Organizations or units • Functions • Experts • Information • Analysis	**People skills** • Building relationships • Interaction skills • Communication skills

SOURCE: Scott et al. (2007).

maintained that common Army planning skills, associated with the mastery of military decisionmaking skills, were essential and largely sufficient for mastering JIIM contexts. Moreover, "delegating to the right talent" sounds a lot like the approach of many senior leaders to coping with the multinational context. Finally, our respondents found people skills, especially in building and maintaining relationships, to be the very foundation of success in all JIIM contexts.[8]

The similarity between these "compensating competencies" and the KSAs we have associated with the various JIIM domains probably stems from a similarity in situations. Like the Air Force general officers struggling to master an unfamiliar organization, Army officers are struggling to master an unfamiliar operational context. Just like Air Force generals, Army officers may not immediately understand the dynamics of their area of operations, but it helps to know what the United States is trying to achieve in a given operation and what the Army's role in that effort is. Just as Air Force general officers approach problems by ensuring that the right people collaborate, Army officers ensure that other agencies are represented in planning and assessment venues. And, of course, the foundation of the military decisionmaking process is a thorough understanding of the problem. Finally, as in any collaborative enterprise, people skills are required to reduce the friction inherent in differing worldviews and differing objectives.

This similarity would seem to recommend further investigation of the application of "compensating competencies" to the JIIM context. But it also recommends a certain degree of circumspection. As *Compensating for Incomplete Domain Knowledge* establishes, "compensating competencies" enable officers to perform satisfactorily in unfamiliar contexts. Those who have developed domain knowledge already, however, usually perform better than those who are compensating. Similarly, the KSAs we have identified may simply be those general competencies that enable officers to cope with unfamiliar contexts.

---

[8] Our respondents' emphasis on interpersonal skills also highlights the relevance of the concept of "emotional intelligence," as described by Daniel Goleman in *Emotional Intelligence*, New York: Bantam Books, 1995, particularly his emphasis on empathy and social skills.

True expertise may very well require a more extensive and more highly developed range of KSAs.

Our analysis should be treated with some caution. Detailed identification and specification of the full range of KSAs associated with each JIIM domain and each echelon, as well as establishing their relative importance, will require further research. In this sense, a key ancillary contribution of our findings is that they can be a framework for further study and analysis.

We can nevertheless draw some broad conclusions with reasonable confidence. First, no single knowledge, skill, or ability appears to be critical for effectiveness in any JIIM context. Overall, the quality and functional expertise of an officer appear to have greater weight in successful performance than any knowledge, skill, or ability unique to any of the JIIM domains. Second, each JIIM domain is in fact cognitively distinct from the others, meaning that developing thorough expertise in that domain requires focused education and significant experience within that domain. Third, the strategic, operational, tactical, and institutional areas appear to comprise distinct domains as well, though their precise outlines are less clear. Fourth, while domain knowledge is important, it assumes this importance mostly at the level of colonel and higher. Below that rank, officers can function effectively in JIIM contexts without specialized JIIM knowledge. Finally, people skills are probably the most important element of success in any of the JIIM domains, just as they are in a broader sense. In short, if an officer is expert in his or her branch or functional area, willing to listen to other perspectives and able to integrate outside input, and able to integrate knowledge and insights from these perspectives in a logical way, there is every reason to believe he or she will be effective (even if not actually expert) in a JIIM context.

CHAPTER THREE

Identifying Developmental Opportunities

Introduction

Our next task was to identify the range of opportunities that developed the KSAs identified in Chapter Two. We focused on assignments, because of the very heavy role experience plays in developing proficiency. Moreover, assignments are the principal focus of our sponsor, the Army Human Resources Command; they are, not coincidentally, the command's principal available developmental tool. That is not to derogate the importance of professional military education in developing proficiency in the JIIM domains, and we encountered several very promising initiatives and proposals in this area.

Our interviews and focus groups naturally elicited considerable information about developmental experiences. That relatively small population did not include the full range of possible Army officer positions, so we supplemented that view by surveying assignment officers at the U.S. Army Human Resources Command and in the Department of the Army's Senior Leader Development Office. Besides the officers currently holding these positions, most of whom are deeply involved in conducting or preparing for combat operations, assignment officers know more about the requirements of the various positions than anyone else. Moreover, we were able to meet with them in order to facilitate their understanding of the KSAs in question and the survey instrument. We did not, however, include reserve component assignment officers in this survey. Further study is therefore required to identify the developmental opportunities available to reserve component officers.

We asked these assignment officers to assess the degree to which the kinds of positions they managed strongly developed any of the KSAs we had identified. To ensure a common understanding of the KSAs in question, we gave a list of definitions to potential survey respondents. We did not, however, explicitly associate any area of knowledge, skill, or ability with a particular JIIM domain. We then analyzed these assessments using criteria derived from the alignment of KSAs with each JIIM domain described in Chapter Two. We sought in this way to determine whether any given type of position conferred primarily joint, interagency, intergovernmental, or multinational experience.[1] In fact, several types of positions appear to develop KSAs that would be useful in several domains simultaneously, especially for deployed soldiers. We reviewed this feedback in light of what our respondents had told us about development.

Finally, we derived the number of actual positions associated with each type of position using the current Personnel Management Authorization Document (PMAD), in order to estimate the distribution of opportunities among the various branches and areas of concentration at each rank. In several cases, these estimates are somewhat imprecise, because we had to further estimate the proportion of officers at those ranks and in those positions who were likely to be deployed.

What we found confirms anecdotal evidence that officers are required to integrate joint, interagency, intergovernmental, and multinational capabilities and concerns even at tactical echelons. Respondents indicated that battalion and brigade commanders, deputy commanders, executive officers, and operations officers are having to address joint and multinational issues routinely. Division and higher-echelon staff officers and commanders must function in a context that is simultaneously joint, interagency, and multinational. Interagency opportunities are otherwise relatively limited, perhaps because of the scarcity of

---

[1] This may seem unnecessarily redundant, at least with regard to the joint domain, given the existence of the Joint Duty Assignment List (JDAL). Earlier RAND studies on the JDAL by Harrell et al. (1996) indicate that the JDAL did not necessarily include every position that conferred joint experience. Moreover, some positions on the JDAL did not seem to actually confer that experience. Also, our interviews indicated that officers were gaining JIIM experience in a much wider range of assignments than included in the contemporary JDAL.

personnel from other government agencies relative to the hundreds of thousands of soldiers and Defense Department contractors taking part in operations. Few of the assignment officers we surveyed indicated that positions they managed below the division level provided much interagency experience. We should also note that assignment officers indicated that positions conferred meaningful joint, interagency, or multinational experience only when the incumbent was deployed on operations. Intergovernmental opportunities are even more restricted for active component officers; few positions were identified as providing robust intergovernmental experience. This makes sense because civil support operations conducted by active component formations are relatively rare, especially when compared with today's contingency operations. Assignment officers did not indicate that lieutenants and captains and officers serving in the generating force were obtaining meaningful experience in any of the JIIM domains.

Finally, our most important finding may be that broadening experiences are crucial to preparing officers for JIIM contexts. Officers usually find such environments new and unfamiliar; success depends on learning from others and enlisting collaboration from individuals and organizations over which one has no formal authority. Our respondents enumerated a variety of experiences that prepared them to operate in one of the JIIM domains, many of which were connected with the domain and echelon in question tangentially, at best. Any job that forces officers out of a narrow focus on their branch or functional area thus makes a major contribution to developing key KSAs in the JIIM domains.

Interview and Focus Group Data: An Imperative for Broadening

Most obviously, respondents noted that experience in a JIIM domain developed KSAs appropriate to that domain. As noted, several of our respondents also contended that other broadening experiences prepared them for working in a JIIM context. General Mike Ryan, the operations officer at the Allied Rapid Reaction Corps, described how growing up in the Far East made him keenly aware of cultural nuances and differing worldviews. General David Fastabend told us that the "hyper-collaboration" he learned in domestic assignments with the Corps of Engineers was a key contributor to his preparation to serve as the Director of Operations at MNF-I. Colonel (ret.) Tony Harriman explained how interacting with the Serbs as a squadron commander in Bosnia was what made him ready for service at the strategic, interagency level on the National Security Council. For Colonel Sean MacFarland, the experiences in cooperation and collaboration with Army organizations he gained as the V Corps Director for Operations (G-3) prepared him for brigade command in the complex joint, interagency, and multinational environment he found in Ramadi. All four of our respondents with formal teaching experience listed that as key to their development (Hoffman, Lamm, Goss, Toczek).[2] At least twenty of the individuals we interviewed identified an experience outside the relevant JIIM domain as important to preparing them to function in that domain; still more supported the concept of broadening experience in some form or other. Some officers, like Mark Quantock, Mo Morrison, and Jay Christensen, did in fact feel that standard development patterns prepared them for JIIM domains. Morrison and Quantock were quick to note, however, that their branch had important joint and interagency aspects starting early in officers' careers.

While these broadening experiences may not have aligned closely with the JIIM domains in their details, they resembled the JIIM context in their essentials. It is not enough that a broadening experience be "different"; respondents described several key aspects of such experiences. They presented officers with new and different situations that they could not master by simply relying on past experience and knowledge, skills, and abilities specific to their branch or functional area. Success required engaging with individuals from different organizational or national cultures, and securing their cooperation without the

[2] Interestingly enough, response from our assignment officer question was about evenly divided, with about half of respondents identifying teaching experience as strongly developing KSAs relevant to one or more of the JIIM domains. It must be noted, however, that service as an assignment officer tends to preclude service as an instructor, meaning that assignment officers have little direct knowledge of the KSAs to be acquired in this job.

support of directive authority. Often, the tasks in question were sufficiently complex that collaboration was essential to developing a solution, not just implementing it.

For many officers, their initial experience in a JIIM context presents similar challenges. They must confront novel and unfamiliar problems and solve them by collaborating with individuals from other organizational and national cultures. Most importantly, such collaboration depends on far more than simply convincing stakeholders to go along with an obvious, U.S.-style solution. In both these developmental experiences and in their JIIM context, officers simply could not solve the problems with which they were confronted without the knowledge, skills, and perspectives that other stakeholders brought to the table.

It may not be possible to provide every officer with experience in one of the JIIM domains. It is probably more feasible, however, to ensure that officers get broadening experiences earlier in their careers. Teaching experience, service on higher-level staffs, and even assignments in the Army's generating force all contribute to developing competencies that will enable Army officers to cope with the JIIM domains' inherent complexity and novelty.

Assignment Officer Survey

Though our respondents provided us with rich information about the kinds of developmental experiences available and the nature of those experiences, our population of around 100 individuals could not conceivably provide useful input as to the relative utility of over 1,000 different kinds of possible positions in developing JIIM KSAs based on their own personal experience. We needed to identify the full range of positions that developed officers for the separate JIIM domains.

To obtain these assessments we turned to a group of officers with recent experience in officer assignments. Our survey population consisted of majors and lieutenant colonels, with between ten and twenty years of experience in their branch or area of concentration. We received responses from eighteen branches and functional areas. This is a relatively small sample. We did not get multiple responses from within a single branch or functional area, and were therefore compelled to aggregate specific positions into categories. For instance, at battalion and higher echelons, we considered commander, deputy commanders, executive officers, and operations officers as part of a "command group." By pursuing this strategy of aggregation, we were able to identify outliers, but also positions about which there was relatively strong agreement.

Data Elicitation

We asked these officers to assess the degree to which various types of positions developed each of the 44 knowledge categories, skills, and abilities we had previously identified as contributing to effective performance in the JIIM domains, using a 5-point Likert scale. Respondents were to identify the kinds of position that "strongly developed" these KSAs. We derived the kinds of positions from those listed in Department of the Army Pamphlet (DA PAM) 600-3, *Commissioned Officer Professional Development and Career Management*. The criteria for determining whether or not a given position "strongly developed" a knowledge, skill, or ability were inherently subjective. We told respondents, however, that they should consider a knowledge, skill, or ability to be "strongly developed" in a given position if an officer could not perform his duties satisfactorily without it. Alternatively, we told that a position "strongly developed" a knowledge, skill, or ability if an officer had to demonstrate superior performance therein in order to receive a superior career evaluation.[3] Respondents recorded their assessments on a modified Microsoft Excel spreadsheet, partially represented by Figure 3.1. They could add remarks further describing the population in question, and also add additional categories of positions not listed on the form.

---

[3] In practice, these two guidelines describe fundamentally different populations. The first category includes those positions analogous to "joint critical" positions on the JDAL. "Joint critical" billets are deemed to require prior experience and to use skills that have already been developed. The second category of position is more developmental in nature and allows incumbents to learn on the job.

Figure 3.1
JIIM Assessment Form

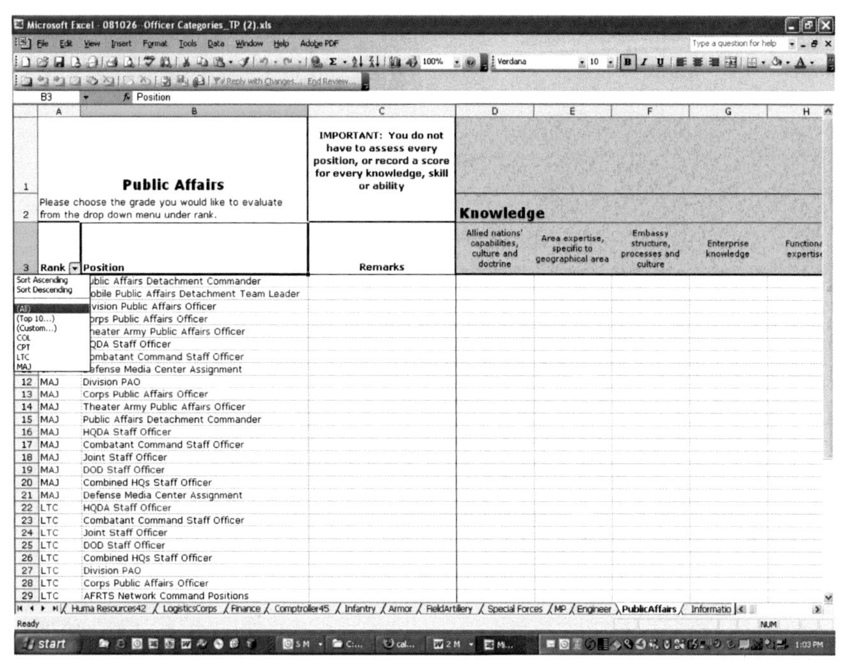

Microsoft Excel - 081026 - Officer Categories_TP (2).xls

Public Affairs

Please choose the grade you would like to evaluate from the drop down menu under rank.

IMPORTANT: You do not have to assess every position, or record a score for every knowledge, skill or ability

Knowledge

	Rank	Position	Remarks	Allied nations' capabilities, culture and doctrine	Area expertise, specific to geographical area	Embassy structure, processes and culture	Enterprise knowledge	Functiona expertise
4		ublic Affairs Detachment Commander						
5		obile Public Affairs Detachment Team Leader						
6		vision Public Affairs Officer						
7		orps Public Affairs Officer						
8		heater Army Public Affairs Officer						
9		QDA Staff Officer						
10		ombatant Command Staff Officer						
11		efense Media Center Assignment						
12	MAJ	Division PAO						
13	MAJ	Corps Public Affairs Officer						
14	MAJ	Theater Army Public Affairs Officer						
15	MAJ	Public Affairs Detachment Commander						
16	MAJ	HQDA Staff Officer						
17	MAJ	Combatant Command Staff Officer						
18	MAJ	Joint Staff Officer						
19	MAJ	DOD Staff Officer						
20	MAJ	Combined HQs Staff Officer						
21	MAJ	Defense Media Center Assignment						
22	LTC	HQDA Staff Officer						
23	LTC	Combatant Command Staff Officer						
24	LTC	Joint Staff Officer						
25	LTC	DOD Staff Officer						
26	LTC	Combined HQs Staff Officer						
27	LTC	Division PAO						
28	LTC	Corps Public Affairs Officer						
29	LTC	AFRTS Network Command Positions						

Sort Ascending
Sort Descending
(All)
(Top 10...)
(Custom...)
COL
CPT
LTC
MAJ

Huma Resources42 / LogisticsCorps / Finance / Comptroller45 / Infantry / Armor / FieldArtillery / Special Forces / MP / Engineer / PublicAffairs / Informatio

RAND *MG990-3.1*

Before administering the survey, the principal investigator briefed potential respondents on the survey's objectives and the study's general findings to date, and demonstrated the use of the survey form. We were able to obtain assessments on about half the branches and functional areas. While we received input from all three functional categories (maneuver, fires, and effects; operations support; and force sustainment), our input was heavily weighted toward the first.

Analysis

Upon completion of the survey, we analyzed respondents' assessments against the KSAs strongly associated with each of the JIIM domains.[4]

---

[4] Again, it is important to remember that these distinguishing knowledge domains, skills, and abilities comprise only a subset of those that contribute to effective performance in a JIIM context.

(See Table 2.1.) It did not seem reasonable to expect that a position should develop all of the KSAs associated with a given domain in order to categorize it as joint, interagency, intergovernmental, or multinational. Accordingly, we developed a set of criteria that allowed positions to be categorized based on subsets of the KSAs associated with each JIIM domain. Note that the qualitative sample that helped guide our criteria was designed with a purposive sampling strategy to illuminate differences and commonalities across domains, echelons, and functional areas with due consideration of context, rather than to provide a sample permitting reliable statistical tests of population data fully generalizable across that population (a goal which, given the complex interaction of individuals, billets, and context, would be difficult to attain). The set of logical associations developed between various KSAs and JIIM domains nonetheless was based on a sample designed to be sufficiently broad in terms of content coverage, and our conclusions were drawn in such a way as to maximize the stability of our findings. Given our relatively small sample, it is important that we explicitly identify the assumptions upon which the criteria are based.

We classified positions *joint* if they developed KSAs according to the following criteria:

- The position had to develop understanding of other U.S. services' capabilities, culture, and doctrine. Of the 241 observations pertaining to the joint domain, 47 identified the importance of this knowledge. This was three times the number of observations pertaining to any other knowledge, skill, or ability associated with this domain. Moreover, integrating other service capabilities is the essence of "jointness."
- The position had to develop one of the following categories of knowledge or skills: understanding joint capabilities and doctrine, understanding joint organizations and processes, managing joint planning processes and systems, or actually employing joint capabilities. We had defined "joint" capabilities, doctrine, etc., as those areas or capabilities that were unique to formal joint contexts, e.g., joint doctrine or the Joint Capability Integration and Development System (JCIDS). We required the position to

meet only one of those criteria because we could easily imagine joint positions whose requirements differed significantly by echelon. For instance, tactical commanders would probably have to understand the employment of joint capabilities, while combatant command planners would also have to be able to use the joint planning process.

- Given the importance associated with understanding one's own organization's capabilities in a collaborative context, the position had to develop knowledge of either "U.S. Army capabilities, culture and doctrine" or "U.S. Army organization and processes."
- It had to develop either "people skills/building and maintaining relationships" or "negotiation/conflict resolution/persuading/influencing," because our respondents identified these skills as indispensable in JIIM contexts. In fact, this criterion applied to all four domains.

Interagency positions fit a different set of criteria:

- Similar to joint billets, an interagency position had to develop understanding of other U.S. government agencies' capabilities, culture, and processes.
- To qualify, the position had to strongly develop two of the following four knowledge areas: "general cultural understanding," "U.S. government budget policy and processes," "U.S. government strategy and policy," and "embassy structure, processes, and culture."
- As with all other domains, a position had to develop either "people skills/building and maintaining relationships" or "negotiation/conflict resolution/persuading/influencing."

Because we had been unable to conduct many interviews on the *intergovernmental* domain, we developed relatively unrestricted criteria. A position qualified in this category if it:

- Developed *one* of the following: understanding the National Incident Management System (NIMS), the statutory and regulatory

environment for Homeland Defense, personal acquaintance with U.S. state and local officials, or understanding U.S. state governments' varying organization and structure.
- Required knowledge of U.S. Army capabilities, culture, and doctrine, knowledge of U.S. Army organization and processes, skill in the military decisionmaking process and planning, or the joint operations planning process.
- Helped to engender either "people skills/building and maintaining relationships" or "negotiation/conflict resolution/persuading/influencing."

Finally, a position developed knowledge, skills, and/or abilities in the *multinational* domain if it required:

- Incumbents to know allied nations' capabilities, culture, and doctrine; NATO capabilities, organization, policy, doctrine, and processes; or international partners' capabilities, culture, and processes.
- "Area expertise," "international/nongovernmental organizations' capabilities, culture and processes," or a well-developed background in other nations' and organizations' theory and doctrine for stability operations and counterinsurgency operations.
- Skills in "active learning/self-initiated learning," "project management, including change management," or "employing Army capabilities."[5]
- "People skills/building and maintaining relationships," "negotiation/conflict resolution/persuading/influencing," "coordinating with personnel from other U.S. organizations," or "foreign language skills." Again, each of these seemed to pertain to applying interpersonal skills in a multinational setting.
- Ability in at least one of the following: "critical thinking/judgment and decision making," "management of personnel resources," or "deductive and inductive reasoning."

---

[5] We grouped "employing Army capabilities" in this criterion because it did not seem essential in and of itself, but fit nowhere else.

To compile Table 3.1, we aggregated the various positions described in each branch and functional area's chapter in DA PAM 600-3 into the broad categories seen in the lefthand column. We then totaled the number of separate respondents who had judged that one or more positions in these categories strongly developed KSAs in a particular domain. For instance, the number "8" in the "Joint" column and the battalion command group row indicates that eight separate respondents thought that service as a battalion commander, operations officer, or executive officer strongly developed joint KSAs. We used the number of responses as the basis for our level of confidence (low, medium, or high) that a category of position provided meaningful JIIM experience.

Findings

Assignment officers' assessments indicated that Army officers are gaining joint and multinational experience at much lower echelons than those assumed by the original Goldwater-Nichols legislation. For the most part, billets on the joint duty assignment list (JDAL) are on the Joint Staff, combatant command staffs, or those of other defense activities. But respondents to our survey indicated that service as a battalion operations officer, executive officer, or commander strongly developed KSAs related to those domains. The same could be said of duty as a brigade commander, deputy brigade commander, executive officer, or operations officer. Respondents also indicated that service on a division or corps staff strongly developed KSAs related to each of the JIIM domains.[6]

As can be seen in Table 3.1, experience in the contemporary operating environment develops knowledge, skills, and abilities most strongly in the joint and multinational domains. Assignment officers reported significantly fewer opportunities to develop competencies

---

[6] These assessments are somewhat anomalous, since few division headquarters are frequently involved in intergovernmental contexts in the current operating environment, and it is impossible to be simultaneously involved in a multinational and intergovernmental context.

Table 3.1
Relative JIIM Opportunity

	Joint	Inter-agency	Intergovern-mental	Multi-national
Platoon Leader/Company Executive Officer	2	3	2	2
Battalion Staff	3	2	1	3
Company Commander	1	1	1	2
Battalion Command Group (Commander, XO, S-3)	8*	3	3	8*
Brigade Staff	6	3	2	6
Brigade Command Group (Commander, Deputy Commander, XO, S-3)	12	7	6	13
Division/Corps Staff	11	10	8	10
Joint Staff	13	11	8	14
AC/RC	3	1	2	3
Combat Developer	4	2	2	3
Advisor and Multinational Staff Officer	4	3	2	6
Instructor	7	2	3	5
Generating Force	5	2	2	3
Maneuver Combat Training Center Observer/Controller	4	2	3	4
Battle Command Training Program Observer/Trainer	7	1		5
Other	1	1	1	

*Only nine branches with command opportunity responded to this survey.

NOTE: Red (1–5) = low confidence; yellow (6–9) = medium confidence; green (≥ 10) = high confidence.

associated with the intergovernmental and interagency domains. U.S. units conduct relatively few civil support operations, of which most last at most a few months. Obviously, Army National Guard officers accrue considerable experience in the intergovernmental domain. Officers assigned to the U.S. Army Corps of Engineers comprise another notable exception. As Major General David Fastabend noted, officers in Corps of Engineers divisions and districts must develop a propensity for "hyper-collaboration," as well as a keen sensitivity to local circumstances.

Experience in collaborating with other U.S. agencies, however, tends to be restricted to higher-echelon staffs and commanders and to civil affairs personnel. This circumstance obtains in no small part because of the relative scarcity of those officials. For example, there were just over 300 Foreign Service Officers in Iraq in 2009 (Holmes, 2009). While over 12,000 contractors augmented these officials, their combined strengths were dwarfed by the nearly 400,000 military personnel and contractors in theater (GAO, 2008). The brigade echelon is probably the lowest organizational level at which officers gain experience in working with other government agencies, a task frequently delegated to the deputy brigade commander (MacFarland, 2008; Cheek, 2008; Boden, 2008).

Evidence of meaningful JIIM experience in company-grade positions seems weak. While a few respondents felt that service as platoon leaders and company commanders did provide such experience, most did not. This result does not, by itself, indicate that company-grade officers are not receiving such experience. Further study, perhaps in the form of a large-scale survey, would be required to confirm or deny that hypothesis.

These survey results are very much a factor of the Army's current operating environment, and assume its continuation. This assumption is not unreasonable, and in fact is reflected in joint and Army documents. JFCOM's *Joint Operating Environment* notes that "One cannot rule out the possibility that U.S. military forces will be engaged in persistent conflict over the next quarter century" (U.S. Joint Forces Command, 2008). The 2009 Army Posture Statement predicts, "Looking ahead, we see an era of persistent conflict—protracted confrontation among state, non-state, and individual actors" (Geren and Casey, 2009). As long as this set of conditions persists, it seems reasonable to assume that the *kinds* of positions indicated in Table 3.1 will be able to develop officers' KSAs in the JIIM domains.

What is unclear, however, is the proportion of these positions that will actually provide such opportunities. The extent to which officers in "Army" billets gain experience in the JIIM domains depends almost entirely on the operational circumstances. Almost all of the responses indicating that officers working at battalion, brigade, and even division echelons acquired meaningful JIIM experience qualified the assessment with the remark "when deployed."[7] Consequently, it cannot be assumed that service in any of the aforementioned positions confers meaningful JIIM experience.

Instead, continued reliance on the new Joint Qualification System seems appropriate. The Department of Defense instituted the JQS when it became apparent that officers were gaining joint experience in many other billets besides those listed on the JDAL in the post-9/11 operating environment. Under JQS, officers who feel they have acquired "joint" experience in non-JDAL billets apply for constructive credit, accompanied by a justification describing the nature of their joint experience. A joint board then adjudicates the application and awards credit accordingly. The JQS does not differentiate among the various JIIM domains, however. Therefore, the results we have discussed above allow career managers to determine the nature of an individual's "joint" experience for purposes of developing capability in one of the JIIM domains and for assigning officers to positions appropriate to their experience. For instance, "joint" credit as a battalion commander can be reasonably expected to provide joint and multinational experience, while command of a provincial reconstruction team provides interagency and multinational experience.

Given the uncertainty associated with the future security environment and the fact that obtaining JIIM experience in any given position is highly contingent on the particular context, retaining this system or one like it seems advisable. Our research does not indicate that each and every position of the types indicated in Table 3.1 confers meaningful experience in a JIIM domain, or that officers in those positions acquire such experience under all conditions. It is therefore much more practical to award JIIM credit after the fact, based on the particular circumstances of an assignment, than to attempt to predesignate certain positions as "JIIM-qualifying."

---

[7] Which we would generalize to "when operational circumstances convey it." The more general wording allows for the possibility that some JIIM experiences could be acquired without deployment.

Education and Internships

During the course of our interviews, respondents asserted that the primary benefit provided by joint professional military education was exposure to other services', nations', and agencies' cultures and mindsets. At one of our Army War College focus groups, one respondent asserted that the primary benefit he gained from attending the Command and General Staff School was "sitting next to a Dutch guy" for most of the course. Similarly, respondents who had attended joint courses at the National Defense University identified exposure to other cultures as the most significant contribution to KSAs in the joint and interagency domains. This finding aligns with the importance respondents attached to relationship building and people skills.

We also identified proposals and initiatives that promise intensive development in one or more of the JIIM domains. Because these initiatives are both novel and take place mainly within the framework of military and civilian education, we were not able to elicit assessments of their utility in developing KSAs in the JIIM domains. That having been said, the literature on expertise indicates that education is fundamental to the development of expertise in any particular field of endeavor.

Executive Order 13434 requires the Secretary of Defense to "issue rules or guidance on professional development programs for Department of Defense military personnel, including interagency and intergovernmental assignments and fellowship opportunities." Two Army initiatives address that requirement. The Command and General Staff College's Interagency Exchange Program and the Army National Guard's proposal to establish intergovernmental internships seem to offer considerable promise. The two programs have significantly different foci. The former aims to improve Army officers' understanding of the capabilities and culture of other U.S. federal government agencies. The latter focuses more on the practical details of municipal and regional governance. Both, however, share the objective of improving Army officers' capability to facilitate unified action in complex, contingency environments.

The nascent Interagency Exchange Program builds on joint and interagency elements in Army intermediate level education's common core. Officers are assigned to the national capital region for two years. During the first year, officers receive common core instruction at the Command and General Staff College's satellite campus at Fort Belvoir, Virginia. Afterwards, they serve on the staff of another government agency for approximately one year, to be followed by another year on the Army Staff. This last year allows the Army to capitalize on the officer's interagency experience (Doughty, 2008).

There have also been significant changes in the Senior Service Colleges' student populations. A much higher proportion of students at the various U.S. war colleges come from other services than was the case previously. A much higher number of foreign students are attending as well. As our respondents noted that spending time with students from other cultures, both national and organizational, was usually the most broadening aspect of "joint" education, these developments will undoubtedly improve "joint" and "multinational" aspects of professional military education.

In contrast, the Director, Army National Guard has proposed an initiative focused on providing Army officers with experience in the practical details of governance. The governance issues confronting most deployed Army officers in many ways resemble those encountered at the U.S. state and municipal level more than they resemble business at the federal level. In this initiative, Army officers would spend a year working in state or municipal government. There they would learn not only the practical details of administration, but also how local politics intimately affect that administration. This initiative would have the added effect of acquainting Army officers with state and local officials and constitutional arrangements, which would improve Army capabilities within the intergovernmental domain as well.

There is recent precedent for such a program. Then Major General Peter Chiarelli helped prepare his 1st Cavalry Division for operations in Baghdad in 2005 through collaboration with the cities of Austin and Killeen, Texas (Chiarelli and Michaelis, 2005). Similarly, Colonel Robert Brown helped train the soldiers in his Stryker Brigade for operations in Mosul by having them learn civil administration from the city government in Tacoma, Washington. Lieutenant General Vaughn's proposal would systematize and extend such experience to a wider range of Army officers.

Several respondents suggested a need for "just in time" education and training to prepare officers for service in one of the JIIM domains. Such courses could refresh some officers' existing capabilities, deepen others' knowledge, or provide updates on issues related to a particular JIIM assignment. Certainly the Joint and Combined Warfighting Course prepares officers from all services for their first joint assignment. As we have noted throughout the report, however, there are significant differences between the JIIM knowledge, skills, and abilities required at different echelons. A major destined for the strategy, plans, and policy section on a combatant command's staff probably requires different preparation than a colonel assigned to the joint warfighting analysis section on the Joint Staff. While this effort to identify developmental opportunities did not consider educational options, our findings imply that there is merit in exploring this suggestion further.

Our research on the developmental value of different positions led us to two main conclusions. From our interviews and focus groups, we concluded that the essential attribute of an assignment intended to prepare officers for service in a JIIM context was that it require officers to function in an unfamiliar environment in which they could rely on neither previous experience nor directive authority for success. It was also helpful if that experience forced officers to rely on collaboration with other individuals and organizations pursuing differing agendas to develop solutions and implement them. Obviously, service in a JIIM assignment provided such broadening, but even service in an Army billet, such as on a higher-level staff, could force officers to develop interpersonal and integrating skills critical to success in a JIIM environment.

The second major conclusion we reached was that the amount of opportunities to acquire developmental experience in the various JIIM domains has increased over the last couple of decades, albeit unevenly. In 1987, officers acquired joint experience mostly at operational and strategic echelons, such as at combatant command headquarters, the Joint Staff, or the Office of the Secretary of Defense. Now, officers serving in key staff positions as low as the battalion echelon have similar opportunities to integrate joint and multinational capabilities, *at least when engaged in current operations.* The relative scarcity of personnel from other government agencies seems to limit interagency experience to the division level and higher, while the relative rarity of civil support operations, at least for active component units, seems to restrict the amount of intergovernmental experience available.

As we will discuss in the next chapter, however, experience in JIIM domains may not of itself contribute significantly to the development of proficiency, absent suitable prefatory education in the employment of those capabilities.

CHAPTER FOUR

Developing Army Expertise in the Joint, Interagency, Intergovernmental, and Multinational Domains

Introduction

Having identified and described the knowledge, skills, and abilities in each of the JIIM domains and the assignments that contribute to developing those capabilities, our next step was to construct career models to portray accumulation of experience and the development of proficiency in these domains.

To do so, we had to answer questions like the following:

- What is the nature of "expertise" in the joint, interagency, intergovernmental, and multinational domains?
- How does one attain "expertise" in these domains? What combination of education, experience, and self-study is required to make an officer an "expert?"
- How many experts are required in each domain?
- How many experts can the Army produce in each functional category?

Answering these questions will enable us to determine whether the Army needs to make any significant changes in officer career management patterns, and thus in the policies that govern them, in order to meet anticipated demand.

One question we could not answer, however, concerned the degree of proficiency actually required from Army officers in the JIIM domains. Our research did not address that question. We were not able to ascertain whether the JIIM domains' scope was more like that of branches and functional areas, or like that of additional skills. Moreover, we could find no other source that did. Intuitively, not every position that requires a degree of competence in the JIIM domains requires true expertise, a quality that, according to the academic literature, may take as much as ten years of sustained effort to develop.

For that reason, we modeled the inventory the Army could produce under two basic approaches: one that portrays the maximum distribution of JIIM experiences across the entire officer corps (managing skills), and one that maximizes the number of officers with multiple tours—i.e., deep experience—in a given JIIM domain (managing competencies). In keeping with our general finding that the fullest range of domain knowledge and the best-developed skills and abilities are required at the colonel level, we focused our investigation on that rank. We constrained our modeling to reasonable feasibility bounds, including inventory management and promotion flow considerations. We should also make clear that we limited our analysis to positions available to active component officers. Without having a firm idea of the full range of reserve component positions that developed KSAs in the JIIM domains, we could not assess either the requirement for proficiency or the reserve components' ability to develop and maintain an inventory of experts.

We found that the Army could easily produce enough colonels to fill billets that might require some prior JIIM experience under a managing-skills approach, and could probably produce enough colonels under a managing-competencies approach. The underlying assumption on which a managing-skills approach rests is that the JIIM domains' scope resembles that of an additional skill. We estimated requirements using the number of colonel positions that respondents indicated "strongly developed" KSAs in each JIIM domain. This was very much an upper bound, however, since that total included both positions that developed JIIM KSAs as well as those that *required* them. Moreover, an increased number of civil servants developed as "national security officers" under Executive Order 13434's impetus might reduce the requirement for Army officers with expertise in the JIIM domains.[1]

If, however, the JIIM domains' scope resembles that of a branch or functional area, then a managing-competencies approach would be required. A strictly interpreted and modeled managing-competencies approach would produce significantly fewer "experts" and could not meet the maximum aggregate demand. However, since not every position that develops JIIM knowledge, skills, and abilities actually requires them, and since even those that do require such competencies do not require them to the same degree, the Army could probably meet the demand for experts under a managing-competencies approach as well.

In all probability, the Army will want to adopt some combination of both approaches. Either approach requires deliberate management to ensure that officers with KSAs in the appropriate JIIM domain are matched to billets requiring those competencies. Even with a significantly expanded pool of opportunities to acquire JIIM experience, not even a managing-skills approach can ensure that *all* officers acquire the required JIIM knowledge, skills, and abilities. Assignment officers will have to manage their inventories carefully to ensure that enough high-quality officers acquire the appropriate developmental experiences to enable them to fill JIIM billets.

Theoretical Background

In the introduction to *The Cambridge Handbook of Expertise and Expert Performance,* K. Anders Ericsson defines *expertise* as "the characteristics, skills and knowledge that distinguish experts from novices and less experienced people" (Ericsson, 2006a). Different scholars have advanced different sets of criteria that can be used to identify experts. The following criteria outlined by John Bransford in *How People Learn* (2000) seem to be broadly representative of thinking in this area. (See also Phillips, Klein, and Sieck, 2007; Lord and Maher, 1991.)

---

[1] It might also increase it, as national security officers return to their parent organizations with a greater appreciation of the necessity for close coordination with the military.

1. Experts notice features and meaningful patterns of information that are not noticed by novices.
2. Experts have acquired a great deal of content knowledge that is organized in ways that reflect a deep understanding of their subject matter.
3. Expert knowledge cannot be reduced to sets of isolated facts or propositions, but instead reflects contexts of applicability: that is, the knowledge is "conditionalized" on a set of circumstances.
4. Experts are able to flexibly retrieve important aspects of their knowledge with little additional effort.
5. Though experts know their disciplines thoroughly, this does not guarantee that they are able to teach others.[2]
6. Experts have varying levels of flexibility in their approach to new situations.

Two key points emerge from these six criteria. The first is that domain knowledge is an essential component of expertise. To attain expertise, an individual must master a significant body of factual and theoretical knowledge. This observation is consistent with the evidence provided by our respondents as to the importance of domain knowledge, but contrasts with the assertion from many of our respondents that they required no specialized knowledge to succeed in JIIM domains. But we note also that respondents with more experience and seniority in these domains contested that assertion. Given the weight of scholarly evidence, we tend to side with the latter group.[3]

The second key point is that experts must be able to apply their knowledge in an appropriate context. For example, doctors diagnose illnesses based on a number of environmental cues, many of which are highly contingent and subjective (Norman et al., 2006). In his investigations, Gary Klein has found that expertise is virtually inseparable from context (Klein, 1998). Indeed, in controlled laboratory experi-

---

[2] Pedagogical ability is distinct from understanding of the subject matter; successful teachers must of course have both (Bransford, 2000).

[3] We must note, however, that both groups agreed that functional expertise was far more important for effective job performance than expertise in any of the JIIM domains.

ments focusing on discrete variables and divorced from actual context, recognized experts frequently perform no better and sometimes perform worse than novices (Phillips, Klein, and Sieck, 2007; Ericsson, 2006a, 2006b). The importance of context places a corresponding weight on experience. The nature and extent of experience appear to be the principal determinants of expertise.

The KSAs we described in Chapter Two seem to be part of the context in which practitioners develop and apply their functional expertise, rather than the core of that expertise. For all the extent and variety of these lists, they do not constitute the entirety of what was required to perform the different jobs we investigated, or even the most important part. Most of our respondents affirmed that functional expertise was more important than any specifically JIIM competency. In concrete terms, an Army logistician would probably succeed in a joint logistics position without prior joint experience or education, while an experienced joint operational planner with a maneuver background probably would not succeed as a joint logistician.

At the same time, the various JIIM domains added significantly to the range of available capabilities, processes, and constraints in a given functional domain. For example, while USAID might be able to provide much-needed development assistance in a given area of operations, planners must forecast requirements far in advance in consonance with the annual and supplemental appropriations cycles. According to our more experienced respondents, attaining mastery of these additional capabilities, processes, and constraints required repeated exposure.

Accordingly, we maintain that JIIM expertise consists of applying an officer's functional expertise in a joint, interagency, intergovernmental, and/or multinational context. This expertise includes understanding the tactical, operational, or strategic effects of joint, other services', agencies', and nations' capabilities and how those capabilities complement Army capabilities within a given functional domain. It also includes an ability to manipulate the associated processes in order to leverage and employ the capabilities in question. Actual application is critical in developing expertise, in that only through application can officers experience the feedback loop that associates action and result. This feedback loop also reinforces understanding of context.

Context is important for another reason. As we discussed in Chapter Two, the descriptions of KSAs in each of the JIIM domains remain relatively stable across echelons and across functional categories. Their actual application differs significantly across these differing contexts in ways that defy precise description, however. For example, officers at different echelons experience considerably different environmental cues. Tactical commanders can base their responses on physical stimuli, such as sight, sound, and body language of the people with whom they are dealing. Operational-level staff officers, however, must base their assessments mostly upon reports and analyses, making for two distinctly different analysis and feedback processes. In concrete terms, while a PRT commander might describe substantially the same set of JIIM knowledge, skills, and abilities as the U.S. mission's comptroller, their actual jobs differ substantially. Developing JIIM expertise thus would seem to require repeated experience within a given echelon and functional category, as well as within the appropriate JIIM domain.

From our study's perspective, the degree of experience is as important as the kind of experience officers get. The literature on the subject is hardly definitive, owing in no small part to difficulties in defining and identifying "experts" to study (Phillips, Klein, and Sieck, 2007). Respondents to a major survey on joint experience indicated that developing relative mastery, as opposed to a threshold level of competence, took about two to three years (Thie et al., 2005). Nevertheless, as we mentioned above, most scholars seem to agree that developing expertise in a given domain requires about ten years of sustained effort. Psychologists initially derived this figure from the study of manual telegraphy and chess, but it has held up in studies of medicine, music, sports, and military decisionmaking. In most of these fields, experts honed their knowledge, skills, and abilities through almost daily application and practice of a narrow range of activities over a decade (Ericsson, 2006a, 2006b; Ericsson, Krampe, and Tesch-Römer, 1993; Lord and Maher, 1991; Bransford, 2000; Norman et al., 2006).

There are some indications that it is the range of experience, rather than its depth, that plays the major role in developing expertise. In 1995 and 1998, Sabine Sonnentag of the University of Amsterdam conducted studies of Swiss and German software-development professionals. In these studies, Sonnentag selected highly skilled professionals on the basis of peer nominations, then analyzed the nature and extent of their work experience. She found no significant difference in length of experience between the moderately skilled (mean of 7.8 years) and highly skilled professionals (mean of 6.6 years). In fact, these figures tell us, if anything, that the highly skilled professionals tended to have *less* actual experience. But the highly skilled professionals tended to have a much greater *variety* of experience, and had worked on over twice as many projects (Sonnentag, 2001). We should note, however, that even if the range of experience outweighs its extent in importance, it still takes time to acquire and integrate that range of experience, between six and seven years in these cases.

We should note also that the relevant experience is not simply a matter of "being there." Rather, expertise results from what K. Anders Ericsson refers to as deliberate practice. At the risk of oversimplifying, we offer that deliberate practice is experience on which a practitioner reflects systematically, usually monitored and guided by a more accomplished practitioner, in order to improve future performance. In fact, the Army's training doctrine implicitly incorporates this emphasis on deliberate practice in the form of prescribing training evaluations and after action reviews (FM 7-0, 2008). According to Ericsson, the amount of such deliberate practice is the principal determinant of expertise (Ericsson, Krampe, and Tesch-Römer, 1993).

If we apply the ten-year rule, few of the individuals we interviewed had accumulated enough experience to be considered experts. Few officers had worked ten years in a single JIIM domain, much less in a single JIIM domain at the same echelon and within the same functional category. Observers with extensive experience in joint, interagency, intergovernmental, or multinational matters observed that repeated tours in these areas made a significant difference in officers' performance. On the other hand, some people take less time to develop expertise than others. Napoleon and his nemesis Arthur Wellesley were both accomplished commanders at the operational level while still in their twenties. Finally, the evidence that expertise in JIIM domains substantially affects an officer's effectiveness is somewhat ambiguous.

Our focus group with officers from Allied Command-Transformation observed that while those U.S. Army officers joining the staff recently had less joint and multinational experience, they nevertheless had more talent overall. When asked which group performed better, those with prior experience or those of higher overall quality, the allied officers could not discriminate between the two. This observation tends to suggest that proficiency in JIIM matters forms a distinctly subordinate aspect of an officer's performance, even in a JIIM context.

Finally, one should not neglect the educational aspect of expertise. Education is a necessary but insufficient component in the development of expertise. The precise relationship between education and expertise is difficult to determine. Studies of medical diagnosticians have found that they make little explicit reference to the basic science they learned in medical school in forming their diagnoses. They rely far more heavily on pattern recognition and other schema developed through years of experience (Norman et al., 2006). In his sociological studies of mental health professionals, Andrew Abbott reached similar conclusions. Indeed, Abbott found that many psychologists, psychiatrists, and nurses were able to draw with equal facility upon their experience regardless of their formal academic training (Abbott, 1988).

At the same time, it is impossible to imagine an expert functioning effectively without extensive formal education, at least an expert in one of the more cognitive domains. The research into medical expertise supports the importance of education indirectly, in that older diagnosticians tend to be less able to detect that a particular pattern is misleading them. There is some indication that professional education tends to provide a valuable lens through which to reflect on experience, especially when professionals can access that education later in their careers (Norman et al., 2006). Education also provides context and background for pattern recognition and hones critical reasoning skills. With regards to the JIIM domains, it seems reasonable to conclude that repeated formal education is important and perhaps essential, but that experience far outweighs education in relative importance.

What We Learned from Our Respondents

The interviews and focus groups we conducted did not resolve the question of whether JIIM domains' scope was more like that of branches and functional areas, or like that of additional skills. Many respondents agreed with the military analysts at the Center for Army Analysis, who insisted that JIIM contexts required little extra in the way of knowledge, skills, and abilities beyond basic Army experience and functional expertise. Others agreed with Ambassador Ron Neumann and Colonel Fitz Lee that success in JIIM contexts required extensive prior experience. One can fairly conclude that the requirement for expertise is highly contingent on the context.

Respondents did tend to agree that prior experience in a JIIM domain made a difference, all other things being equal. All other things are seldom equal, however: note again our previous observation from foreign officers at Allied Command-Transformation, that better U.S. officers tended to function more effectively in their JIIM environment, even if those U.S. officers lacked prior joint or multinational experience. Moreover, from our interviews with officials working at the strategic echelon (Hoffman, Freier), it seemed that experience in a given echelon influenced officers' overall success at least as much as experience in a given domain.

As we already noted in Chapter Three, the most important aspect of development was exposure to "something different." Experiencing "something different" could occur in a branch or Army assignment, as is happening daily in Iraq and Afghanistan. It could also happen in an educational setting. Most officers noted that the most important aspect of joint professional military education was exposure to other organizational cultures, rather than any specific knowledge or skill unique to the JIIM domains.

That having been said, most officers also observed that critical thinking skills were very important for success in JIIM contexts, and that education in critical thinking came too late in their careers. This presents an interesting paradox, in that almost all colleges and universities work hard to inculcate critical thinking. It is one of the principal reasons that the U.S. Military Academy teaches philosophy. The same can be said of the Command and General Staff School, the Army War College, and most of the other military educational institutions. We can only conclude that these efforts simply did not reach many of our respondents at a time when they were receptive to them. It is far from clear, however, whether this is because the efforts are inadequate or because many officers are not receptive to education in critical thinking. It may also be that some are just not critical thinkers.

Analytical Strategy

The research team focused our analysis on the Army's ability to develop and maintain the required inventory of officers with a certain level of proficiency in each of the JIIM domains. We used experience as a loose proxy for the desired level of proficiency, realizing that the relationship among education, experience, and proficiency is complex. First, we had to define alternative approaches for developing officers. Under a managing-skills approach, the Army would develop an elementary level of proficiency in each of the JIIM domains by spreading a modicum of JIIM experience among as many officers as possible. The alternative, the managing-competencies approach, attempts to maximize the number of officers with extensive experience in the JIIM domains. Both approaches will be described in more detail later in this chapter.

Obviously, each of these approaches presents advantages and disadvantages. A managing-competencies approach would limit time available for other developmental opportunities, and might result in curtailment of opportunities previously considered highly desirable, if not essential. A managing-skills approach, on the other hand, would probably obviate the need for such tough choices, but might not actually produce the required degree of expertise. We wish to emphasize that we advocate neither of these particular choices; we simply modeled them in order to estimate the possible inventories of JIIM-qualified officers under the least demanding and the most demanding set of conditions.

Next, we had to estimate a requirement for JIIM-qualified officers, since none currently exists (excepting the Joint Duty Assignment

List, which covers predesignated "joint" positions). As a proxy, we used the data provided by our assignments officers about the number of colonel positions requiring proficiency in one of the joint domains.

Finally, we modeled the alternatives using an officer inventory projection model developed by RAND researchers. The intent of this approach was to assess the feasibility of maintaining a desired reservoir of officers with a certain level of JIIM proficiency under less demanding and more demanding specifications for the degree of proficiency required. Our modeling estimates the number of officers with either JIIM experience (less demanding, under the managing-skills approach) or JIIM expertise (more demanding, under the managing-competencies approach) that could be developed if the Army set out to do so. Although our modeling produces quantitative results, the estimates should not be considered precise. Our analysis can best be interpreted as indicating whether or not the probable inventory is approximately greater than, equal to, or less than the requirement specified. The estimates thus provide insights into the relative feasibility of either approach.

The reader should bear in mind that in simulating different goals for the development of JIIM experience, our modeling simulates an Army career management system that endeavors to achieve those goals. In the managing-skills construct, this does not create very much tension with other possible career management goals, like development of deep functional expertise or accumulation of large amounts of tactical experience. The model shows that these goals could reasonably be achieved while still getting a large number of officers into at least one JIIM assignment. Managing competencies, on the other hand, can easily put so much demand on officers' time that they have too little time remaining for other developmental opportunities. Thus, it may be somewhat unrealistic to simulate accumulating experience at the managing-competencies level: the Army will probably continue to prioritize the development of functional expertise, a preference entirely consistent with this research. Thus, since it is by no means clear that the Army should accord JIIM experience a priority high enough to meet managing-competencies criteria, the managing-competencies estimates at which we arrive should be considered upper bounds: feasible, but most likely not optimal considering the Army's other developmental goals.

Readers should also note that our modeling does not simulate any prioritization among the JIIM domains: each is accorded equal importance, so again the result is a feasible range of possibilities.

Managing Competencies: The "Ten-Year Rule"

Applying the ten-year criterion to Army careers is more complex than it might seem at first. The Army typically manages officers' careers by the number and type of assignments, rather than by their cumulative duration. Complicating things still further, actual assignments tend to vary in length, especially as officers rise in rank. Assignments can last anywhere from one year in a deployed joint task force to three years, the nominal length of a standard Army assignment. Dividing ten years by a three-year tour results in three assignments with a remainder. Given Sonnentag's findings that variety of experience may matter as much as or more than depth, it seemed even more reasonable to look at the number of assignments rather than their cumulative duration. We further reasoned that intermediate-level education and attendance at a senior service college would consume two years and provide the necessary educational component for developing expertise in a JIIM domain. This approach seemed reasonable given recent initiatives to enhance the JIIM components at the Command and General Staff School and to provide Joint Professional Military Education (JPME) Level II at all the senior service colleges. In short, we translated the "ten-year rule" into three assignments in a given domain.[4] The managing-competencies approach consisted of the normal course of military education and three assignments.

It is very likely that the three assignments we specify above do not all have to be in the same JIIM domain. Almost all of our respondents insisted on the primacy of people skills and other more general skills—like communication and reasoning—in functioning successfully in any of the JIIM domains. As a corollary, respondents maintained

---

[4] Three assignments could approach ten years' experience. More likely they will provide somewhere between five and eight.

that effective people skills for operating in a JIIM environment were best developed by assignment to a position outside a person's "comfort zone," requiring mastery of an unfamiliar environment and ability to influence others without being able to exercise directive authority. Positions providing this kind of challenge may be found in any of the domains. For many, that kind of developmental assignment actually occurred for the first time in a JIIM context. For others, like Colonel Sean MacFarland, serving as a corps operations officer provided that experience, in that it involved a far greater scope of activity than others in the typical pattern of Army assignments, without having the same role clarity as positions at the division level and below. The specific nature of the assignment mattered far less than the fact that it was "different" and offered the challenges described above.

Managing Skills: The "One-Tour Rule"

In this case, we assumed that KSAs unique to the JIIM domains were a relatively minor component of the expertise required in those contexts, and that one tour in a JIIM domain should suffice to develop the necessary competencies. As with the "ten-year rule," we assumed that intermediate-level education and senior service school would provide the required education. Because we did not have to consider combining several tours to accumulate long experience, we did not have to address the issue of how well knowledge, skills, and abilities transferred between domains. That observation still applies, however, in the sense that Army personnel managers can probably assign officers who have attained competence in one JIIM domain to positions in another JIIM domain with reasonable confidence that they have acquired enough KSAs to perform effectively, or at least learn to do so in short order. The following analysis estimates how many officers have attained this level of competence in each JIIM domain, by branch and functional area.

The RAND Officer Inventory Projection Model

We used an officer inventory projection model developed for other projects and adapted it to our purposes. The model is a standard, steady-state inventory projection model that optimizes against a given set of criteria. RAND's model accounts for the complexities of modern Army officer management, such as different opportunities and outcomes in different branches and functional areas, the apportionment of branch-immaterial positions, and the various points at which officers can move from a basic branch into a functional area. Its primary use is to compare the effects of different personnel policy options on the composition, experience accumulation, and promotion opportunities of the Army's officer inventory.

Applying this model to the problem of developing JIIM expertise, however, requires the analyst to use considerable care both in structuring the modeling problem and in interpreting the results. Such application was relatively straightforward in the managing-skills approach, since the model relatively easily calculates the degree to which JIIM experience can be distributed. So the model can provide a reasonable estimate of the likely results of policies designed to distribute JIIM experience broadly.

The model did not optimize for any particular JIIM domain, but instead optimized the Army's aggregate pool of experience in all four domains. In other words, the model did not prioritize experience in any domain. It is possible that it might be feasible to trade experience in lower-priority JIIM domains for experience in higher-priority domains. Still, the model provides a reasonable approximation of the number of officers with the requisite experience in specific domains that the Army could produce. Moreover, because many developmental experiences impart experience in several JIIM domains, it is reasonable to argue that optimizing for aggregate JIIM experience will, for all reasonable intents and purposes, optimize for each separate domain as well. If one assumes these officers' assignments are carefully planned to produce domain depth, the results represent an upper bound of the number of experts that could be produced.

The model's results can be confusing, reflecting the underlying reality. As shown in Figure 4.1, the model projects the population of colonels with aggregate JIIM experience who meet a certain set of criteria, e.g., number of assignments. That population contains people with assignments in the four JIIM domains. Just as in reality, many of the people in the model with experience in one domain also have experience in another. And also just as with reality, many of those assignments

Figure 4.1
Interpreting RAND Officer Inventory Projection Model Results

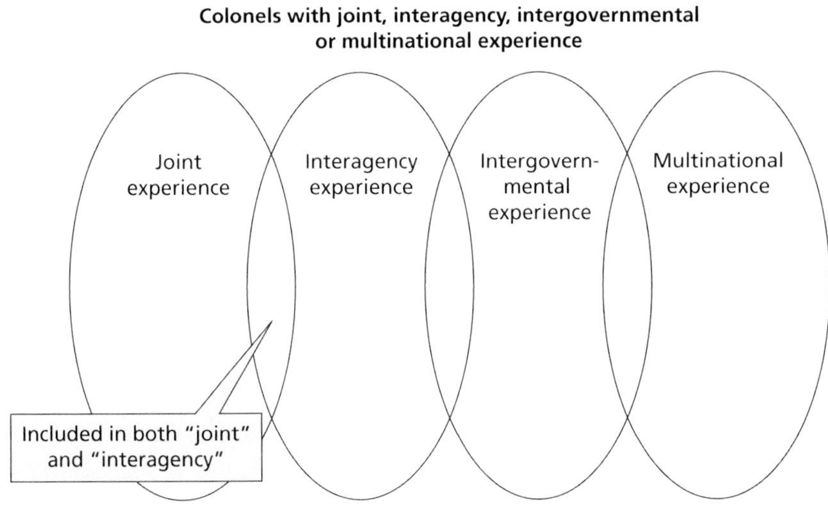

RAND *MG990-4.1*

combine experience in two or more domains. Thus the model's output for the maximum number of colonels with a certain number of joint tours will be the maximum number of colonels who have had the required number of joint assignments, a figure that will include colonels who also have met the desired criteria for experience in one or more of the other domains. Those officers who meet the desired criteria in several domains will be reported in each domain, utilizing a series of tables. But the results depicted address only the question implicit in each table's title, and do not attempt to reconcile the answers into one overarching sum. Readers would thus be well advised to avoid trying to reconcile the tables against one another or internally.

Identifying Potential Requirements

In order to assess the degree to which the Army could produce the number of JIIM experts in our various modeling scenarios, we first had to estimate the number of billets that might require such expertise. Our interviews suggested that incumbents required domain expertise as colonels, so we focused on estimating the number of colonel positions that might possibly require expertise in a given JIIM domain. Reasoning that any position that strongly develops JIIM competencies at the colonel level likely requires some degree of those competencies to start with, we simply added up the number of colonel positions identified by our respondents as strongly developing KSAs associated with each JIIM domain.

Those familiar with the JDAL will note that the number of positions significantly exceeds the current number of "joint critical" positions for colonels, of which there are currently 123. Because our instructions to our respondents specified that "strongly develop" could include positions that developed KSAs without requiring them, as well as those positions that actually required incumbents to demonstrate competence in them, the estimates reflected in Table 4.1 should be considered an upper bound. Moreover, because positions frequently developed KSAs in more than one JIIM domain, the estimate above may overstate the requirement. On the other hand, one reason officers have "increased opportunity" to acquire JIIM experience is that they are more and more often finding themselves in situations that require them to integrate Army activities with those of joint, interagency, intergovernmental, or multinational partners. Positions that give rise to these situations are of course part of the "demand" for JIIM proficiency, but they are also part of the "supply" of opportunities to develop that proficiency. Thus, for all the reasons discussed above, the reader should remember that Table 4.1 represents an approximation, not a precise estimate.

Table 4.1
Maximum Number of Positions Requiring Expertise in Each JIIM Domain

Domain	Number of Positions
Joint	357
Interagency	341
Intergovernmental	171
Multinational	380

Findings

In this section we compare the estimated requirements depicted in Table 4.1 with the results of our inventory modeling of the managing-skills and the managing-competencies approaches. After presenting a broad overview of our findings, this section lays out the results of our modeling for each of the JIIM domains under both approaches. As discussed earlier, our model simulated career management decisions that would maximize the number of officers meeting the experience specification under the approach being modeled.

In general, our modeling indicated that the Army could supply its requirements for officers who are proficient in the various JIIM domains under either a managing-skills or a managing-competencies approach. A managing-skills approach easily would produce more officers with domain experience than the potential requirement indicated in Table 4.1 above. Clearly, the more restrictive criteria inherent in a managing-competencies approach would produce fewer officers with any domain experience than a managing-skills approach, and many fewer officers with three or more domain assignments.

Still, the number of such experts would approach the maximum number potentially required, while the number of officers with at least some domain experience would exceed that requirement. Table 4.2 depicts the number of colonels with at least one assignment in the four JIIM domains. Clearly, the managing-competencies approach would limit opportunity to attain the desired experience, as reflected by the significantly lower figures associated with it in Table 4.2. Still, the results of both approaches comfortably exceed the minimum requirements set forth in Table 4.1, indicating that establishing the more stringent criteria inherent in a managing-competencies approach will not render the Army unable to provide a threshold level of experience in any domain.

Some branches and functional areas have better opportunity than others to acquire the needed experience. Our modeling indicates that officers in the operations support functional category have significantly more opportunity to acquire experience in JIIM domains than officers in either the maneuver, fires, and effects or the force sustainment

Table 4.2
Projected Inventory of Colonels with at Least One Assignment in Each JIIM Domain Under a Managing-Skills and a Managing-Competencies Approach

(Required)	Joint (357)	Inter-agency (341)	Intergovern-mental (171)	Multi-national (380)
Managing skills	1,812	1,287	840	1,765
Managing competencies	1,331	898	660	1,407

NOTE: The numbers do not sum horizontally, since many of the people who have experience in one JIIM domain could also have experience in several others.

functional categories, at least relative to their overall proportion within the officer corps. In retrospect, this circumstance is not all that surprising. The operations support functional category includes the signal and military intelligence branches, along with functional areas such as strategy, plans and policy, strategic intelligence, operations research and systems analysis, and foreign area officers. Officers in these functional areas often serve on higher-echelon staffs, many of which are joint and many more of which afford significant exposure to JIIM experiences. Signal and military intelligence officers also serve on such staffs and, when in other assignments at lower echelons, frequently integrate capabilities from other agencies and services. On the other hand, our modeling indicates that officers in the force sustainment functional category have significantly less JIIM opportunity than either of the other two functional categories, especially relative to their overall representation within the officer corps.

While the Army probably neither can nor should accord developing JIIM capabilities absolute preeminence among other developmental priorities, the results depicted indicate that the Army could adopt either a managing-competencies or a managing-skills approach: the estimates show the Army can produce and maintain the required number of experts in each JIIM domain.

In fact, some combination of the two approaches is probably desirable. As experience in the current operating environment, congressional legislation, and executive branch policy make clear, wider dissemination of proficiency in the JIIM domains is highly desirable.

In many assignments, a basic level of proficiency will suffice. In some positions, however, senior leaders will have to demonstrate true expertise in integrating and synchronizing capabilities and activities in the JIIM domains. The Army could adopt a managing-skills approach for most officers, while employing a managing-competencies approach sufficient to develop and maintain an inventory of experts large enough to fill certain key positions.

Finally, the resulting projections include officers who would have attained experiences in several domains; some of the people with two joint experiences, for example, are the same people who have gained three interagency experiences because two of their assignments were assessed as being both joint and interagency. For that reason, it is not possible to simply add the estimated number of colonels with joint experience to those with interagency experience, and so on, in order to arrive at an estimate of colonels with some sort of JIIM experience. Such a procedure would lead to considerable double counting.

Projected JIIM Inventory Under a Managing-Skills Approach

Assuming that current conditions continue to prevail, the Army will be able to maintain a significant inventory of senior leaders with at least one joint, interagency, intergovernmental, or multinational experience under a managing-skills approach. As shown in Table 4.3, the vast majority of colonels (over 2,000) will have completed at least one assignment in a JIIM context by the time they retire or are promoted. The column for three assignments is empty because we are modeling the widest possible distribution of JIIM assignments. In other words, no officer gets a third assignment unless all have gotten at least one. This result models the Army distributing JIIM experience as widely as possible.

The Army should be able to produce enough officers with joint experience using a managing-skills approach. According to Table 4.1, the Army would need to maintain an absolute maximum 357 joint experts at the colonel level. As we see in Table 4.4, the Army would comfortably exceed that requirement. Indeed, the number of officers with two prior joint assignments is more than double the potential requirement.

Table 4.3
JIIM Experience Accumulated Under a Managing-Skills Approach

	Projected Inventory of Colonels with at Least One Prior Assignment in Any JIIM Domain		
	1	**2**	**3**
Maneuver, fires, and effects	497	428	NA
Operations support	158	283	NA
Force sustainment	403	272	NA
Total	1,059	983	NA

NOTE: Some readers will note the discrepancy between the totals in Tables 4.2 and 4.3. As noted in text, the totals for each JIIM domain indicated in Table 4.2 cannot be aggregated, because those figures include many of the same individuals. The only clue that Table 4.2 provides to the total pool of officers with at least one assignment in any JIIM domain is that at least 1,812 colonels have attained one experience, because that is the greatest number shown. In fact, considered cumulatively, 2,042 colonels can attain at least one experience in one of the JIIM domains, as indicated in Table 4.3. In contrast to Table 4.2, the figures in Table 4.3 should be added to indicate the maximum number of colonels who have obtained one JIIM assignment. Thus Table 4.3 indicates that the Army could produce 2,042 colonels with at least one JIIM assignment under a managing-skills approach. The reader will be well advised not to try to read more into any of the other tables than is expressly indicated by its title.

Table 4.4
Joint Experience Accumulated Under a Managing-Skills Approach

	Projected Inventory of Colonels with 1, 2, or 3 Prior Joint Assignments (357 Required)		
	1	**2**	**3**
Maneuver, fires, and effects	500	332	NA
Operations support	103	253	NA
Force sustainment	428	197	NA
Total	1,031	781	NA

The Army also ought to be able to produce well over the required number of colonels with at least one interagency experience. Table 4.5 indicates that the Army would be able to produce close to 1,300 colonels with some interagency experience, sometime in their career. Officers would have considerably reduced opportunity for repeated tours, however.

Table 4.5
Interagency Experience Accumulated Under a Managing-Skills Approach

	Projected Inventory of Colonels with 1, 2, or 3 Prior Interagency Assignments (341 Required)		
	1	2	3
Maneuver, fires, and effects	400	209	NA
Operations support	237	29	NA
Force sustainment	387	24	NA
Total	1,025	262	NA

Under a managing-skills approach, the Army could produce almost 900 colonels with at least one intergovernmental experience (Table 4.6). While the number of colonels with intergovernmental experience that would be produced is considerably less than the number of colonels with interagency experience, so is the requirement. The maximum number of colonels with intergovernmental experience required is about 170, approximately half of the number possibly required in the interagency domain.

Thus, the Army can produce an adequate number of officers to fill senior leader positions requiring multinational experience. Given the highly multinational nature of the current operating environment and projection that it will persist, the Army probably can meet its requirements if a managing-skills approach is appropriate. As shown in Table 4.7, close to 1,800 colonels can expect to have had at least one significant multinational experience.

Table 4.6
Intergovernmental Experience Accumulated Under a Managing-Skills Approach

	Projected Inventory of Colonels with 1, 2, or 3 Prior Intergovernmental Assignments (171 Required)		
	1	2	3
Maneuver, fires, and effects	337	38	NA
Operations support	208	7	NA
Force sustainment	234	16	NA
Total	779	61	NA

Table 4.7
Multinational Experience Accumulated Under a Managing-Skills Approach

	Projected Inventory of Colonels with 1, 2, or 3 Prior Multinational Assignments (380 Required)		
	1	2	3
Maneuver, fires, and effects	484	261	NA
Operations support	289	106	NA
Force sustainment	388	238	NA
Total	1,161	604	NA

Under a managing-skills approach, then, the Army would exceed the maximum requirement for colonels with at least one JIIM assignment. By seeking to distribute JIIM experience as widely as possible, however, the Army would develop essentially no officers with three or more assignments in a single JIIM domain. If a little experience suffices to prepare officers for a JIIM context, a managing-skills approach would be adequate. If, however, extensive experience is actually required to meet the demands in at least some JIIM contexts, the Army would have to complement a managing-skills approach with targeted development of a cadre of experts, i.e., combine the managing-skills and managing-competencies approaches. Alternatively, the Army could supplement the experience officers had already acquired with "just in time" education to prepare officers for specific assignments.

Projected JIIM Inventory Under a Managing-Competencies Approach

Under a managing-competencies approach, the Army would be able to develop significantly fewer experts, defined as colonels with three or more tours in one domain. Still, as indicated in Table 4.8, the number of such experts usually approaches the maximum number potentially required in each domain as indicated in Table 4.1. The reader should also note that adoption of a managing-competencies approach would still produce a significant number of colonels with one or two JIIM experiences in addition to the quantity of experts with three or more accumulated assignments.

Table 4.8
JIIM Experience Accumulated Under a Managing-Competencies Approach

	Projected Number of Colonels with 1, 2, or 3 Prior Assignments in Any Single JIIM Domain		
	1	2	3
Maneuver, fires, and effects	242	258	199
Operations support	95	132	139
Force sustainment	209	212	145
Total	547	602	484

Under a managing-competencies approach, the Army could maintain nearly 500 colonels who have completed three joint assignments, as shown in Table 4.9, if assumptions about the operating environment and available opportunities remain valid. That number would exceed the potential requirement of 357 depicted in Table 4.1. In addition to the nearly 500 experts, there would also be approximately 1,100 other colonels who would have accumulated at least one joint assignment at some point in their careers.

Under a managing-competencies approach, however, the Army would not be able to produce the maximum possible number of interagency experts required. As Table 4.10 notes, the Army would be able

Table 4.9
Joint Experience Accumulated Under a Managing-Competencies Approach

	Projected Number of Colonels with 1, 2, or 3 Prior Joint Assignments (357 Required)		
	1	2*	3
Maneuver, fires, and effects	162	271	158
Operations support	45	137	125
Force sustainment	87	220	127
Total	293	628	410

* The reader may have noted that numbers in this column exceed the numbers of those officers with two JIIM experiences in Table 4.8. This is because many of the officers projected to have two joint experiences will actually have attained *three* experiences in *another* JIIM domain, because two or more assignments provide experiences in more than one domain, and are thus accounted for in the "3" column of Table 4.8.

Table 4.10
Interagency Experience Accumulated Under a Managing-Competencies Approach

	Projected Number of Colonels with 1, 2, or 3 Prior Interagency Assignments (341 Required)		
	1	2	3
Maneuver, fires, and effects	173	142	102
Operations support	144	36	22
Force sustainment	120	128	29
Total	438	306	154

to develop only 154 colonels with three prior interagency experiences, well short of the 341 potentially required. Still, the number 341 represents the upper bound of the requirement, probably a rather high upper bound. Moreover, Table 4.10 indicates that there would be just over 300 colonels with two interagency assignments. Taken together, these projections indicate that the Army would probably be able to provide the required degree of expertise, especially if the actual requirement (for three assignments) was somewhat less than 341.

Table 4.11 shows that under a managing-competencies approach, the Army could produce 65 colonels with at least three intergovernmental assignments. As was the case with the interagency domain, this would fall short of the 171 required. Still, it is far from clear that the Army would in fact require 171 experts to meet demand. In addition, the fact that 152 colonels would have accumulated two such assignments again bodes well for the Army's ability to produce a sufficient inventory of intergovernmental experts. Finally, our analysis included only those opportunities open to active component officers. The Army could also draw on its cadre of National Guard officers to meet this demand.

Similarly, Table 4.12 shows that the Army, by using a managing-competencies approach, would produce somewhat fewer than the 380 multinational experts nominally required. That requirement includes the fiscal year 2008 addition of 18 colonel billets for advisors and trainers of foreign forces, a number that is likely to grow. For all of the reasons we have described previously, it is likely that the combined inventory

Table 4.11
Intergovernmental Experience Accumulated Under a Managing-Competencies Approach

	Projected Number of Colonels with 1, 2, or 3 Prior Intergovernmental Assignments (171 Required)		
	1	2	3
Maneuver, fires, and effects	253	51	24
Operations support	124	28	14
Force sustainment	66	73	27
Total	443	152	65

Table 4.12
Multinational Experience Accumulated Under a Managing-Competencies Approach

	Projected Number of Colonels with 1, 2, or 3 Prior Multinational Assignments (380 Required)		
	1*	2	3
Maneuver, fires, and effects	292	195	133
Operations support	161	64	77
Force sustainment	165	223	96
Total	618	482	307

* As in Table 4.9, many of the colonels projected to have one multinational experience would be the same officers counted as having two or three JIIM experiences in Table 4.8.

of colonels with the necessary multinational experience would meet the actual demand, however.

In sum, the Army probably would produce significantly fewer officers with some JIIM experience under a managing-competencies approach than under a managing-skills approach. The total number of officers with at least some JIIM experience would still exceed the potential requirement, however. Moreover, the Army would accumulate a substantially greater inventory of officers with deep experience.

Our investigation indicates that under current conditions, the Army should be able to maintain an adequate inventory of officers with the required degree of JIIM knowledge, skills, and abilities, at least if it adopts a managing-skills approach. Adopting a managing-competencies approach, with its emphasis on many varied experiences within a given domain, would pose a greater challenge to the Army, but still one that the Army could probably manage.

Our investigation could not, however, establish definitively which of these two approaches was more appropriate to the current context. It was not clear, for instance, that the JIIM domains constituted unique fields of endeavor in the same way that officers' branches or functional areas do. Further, many officers have functioned superbly in JIIM contexts without extensive prior JIIM experience. On the other hand, interviews with several of our more experienced practitioners tended to support the contention that multiple experiences were more likely to lead to optimal effectiveness in senior JIIM assignments.

The latter observations are consistent with the literature on developing expertise. That literature suggests that extensive, varied experience in a given domain over a period of six to ten years is necessary to developing true expertise. Indeed, the Army's templates for developing branch expertise at the tactical level conform to this pattern. Thus, the best career management solution will probably employ both approaches, recognizing that some officers in JIIM positions must be true experts, while others merely require some developmental experience. Neither type, however, is likely to succeed without adequate educational preparation for service in these complex and challenging contexts.

CHAPTER FIVE

Summary and Conclusions

Introduction

In this project we sought to identify the knowledge, skills, and abilities associated with the various JIIM domains, in order to allow the Army to develop its officers more effectively in those domains. In support of this objective, we also sought to identify the various types of experience that developed such knowledge, skills, and abilities. Finally, we assessed the Army's ability to produce and maintain an inventory of officers with the required qualifications.

This chapter collects and summarizes our major findings, and then describes their principal policy implications. These implications concern the issues with which the Army must grapple, and we describe illustrative policy options to further illuminate those issues. This chapter also lays out directions for further research derived from those implications.

A few cautionary notes are in order. First, the need to elicit and analyze a full range of contextual, qualitative data necessarily limited the sample population in this study. Army officials can repose reasonable confidence in our broad conclusions, not least because those conclusions are consistent with other sources of information. Finer judgments, such as those aligning specific items of JIIM knowledge, skills, or abilities with specific echelons or specific positions, for example, probably require a much larger sample population. We have inserted appropriate caveats throughout the text.

More importantly, while we are reasonably confident that we have identified the KSAs that are important in the various JIIM domains, it was not the purpose of this study to establish the importance of these domains relative to other aspects of full-spectrum operations. Put another way, this study describes the qualities necessary to excel in particular JIIM domains, but reaches no conclusions about whether it is necessary to excel in the JIIM domains in order to excel in full-spectrum operations. Intuitively, however, such competencies seem to be fairly important, given the importance of other agencies' and nations' contributions to full-spectrum operations.

Major Findings and Their Implications

We begin by recapitulating the major findings from our investigation:

- **Successful performance in joint, interagency, or multinational contexts requires the application of highly developed functional expertise to novel situations.** Service and branch or functional area expertise are the foundation of Army officers' utility in JIIM contexts. Officers need to understand their specialty well enough to think beyond an Army context, however. A military policeman must be able to think and act like a policeman; a military engineer must be able to perform as an engineer; and so on.
- **The JIIM domains are qualitatively distinct, though overlapping.** Simply put, success in each of the JIIM domains requires a different set of KSAs; expertise in one JIIM domain does not completely translate to expertise in another. To the extent that actual expertise is required, developing expertise in any one of these domains requires focus and repetitive but varied experience.
- **The strategic, operational, tactical, and institutional echelons require distinctly different KSAs.** That is, jobs at these different echelons differ in kind, not merely in degree. A combatant command staff is not just a bigger, more capable brigade staff.
- **Broadening experiences contribute significantly to competence in the JIIM domains.** For some, that broadening experience was service in the Balkans; for others, it was a tour on a higher-level staff; and for one, it was working with KATUSAs (Koreans

Attached to the U.S. Army). What all these experiences had in common was that they forced officers to cope with an unfamiliar context, and that mission success depended on effective, entirely voluntary cooperation from other individuals and organizations.

- **In the current operating environment, Army officers have significantly increased opportunities to gain experience in one or more JIIM domains.** Our survey of assignment officers indicated that even service in "Army" positions, such as battalion or brigade commander, executive officer, or operations officer, provided significant experience in integrating joint and multinational capabilities. Officers who served on division and higher-echelon staffs accumulated significant interagency experience as well. Development required deployment, however. In a garrison setting, those same positions provided little opportunity for developing JIIM-relevant KSAs.
- **It should be possible to develop and maintain enough officers with the required KSAs in the JIIM domains.** Our modeling showed that if the Army adopted a managing-skills approach, over two-thirds of lieutenant colonels would have had some sort of JIIM experience by the time they either retired or became colonels. All colonels would have accumulated a JIIM assignment sometime in their career. While the Army could produce substantially fewer experts, but with deeper experience, by using a managing-competencies approach, it is likely that the resulting inventory would still satisfy demand. Either approach requires deliberate, effective management, however.

We restate these findings to refresh the reader's memory before discussing their implications. It is worth noting that while this study's purpose was not to recommend changes to the Army's officer personnel management policies, the major findings listed above have significant implications for Army personnel policy.

First, maintaining an adequate reservoir of expertise in each JIIM domain will require deliberate management. While opportunities to acquire experience in these domains have increased, such opportunities are not universal. Over time, particularly if operational tempo declines, these opportunities may decrease. Second, whether or not a given position actually offers experience in one or more JIIM domains is, in many cases, so highly contingent on what the incumbent actually does during his or her assignment as to preclude prior designation as an "interagency" or "multinational" experience. In these cases, at least, Army officials will continue to have to rely on the Joint Qualification System or some future analog in order to determine whether officers have been developed in a JIIM domain. Third, the fact that the JIIM domains are qualitatively different from one another will require assignment officers to track the kind of experience an officer has had—joint, interagency, intergovernmental, or multinational—not just whether he or she has had one such position. The current Joint Qualification System does not make such distinctions. Instead, it aggregates all such JIIM experience under the heading "joint."

Army careers may simply be too short to develop adequate expertise in both the JIIM domains and an officer's basic branch or functional area. The Army takes between fourteen and sixteen or more years to develop battalion commanders, who might reasonably be called its tactical "experts." The most focused development patterns alternate service in tactical units with assignments that are closely related, such as serving as an observer/controller at a combat training center or a small group instructor at a branch school. This is especially significant as a sizable portion of our respondents identified functional expertise as a necessary predicate to success in JIIM contexts. If it does, in fact, require between six and ten years of focused experience to make an officer expert in a JIIM domain, that means developing officers who are both experts in their branch and experts in a JIIM domain could require anywhere from twenty-two to twenty-six years. The current security environment does make some of the experiences simultaneous, e.g., commanding a brigade combat team in a stability operation, and thus can shorten this period. Our respondents were very clear, however, that a brigade commander who was not deployed would not acquire significant joint, interagency, or multinational experience. Under a significantly less demanding operational tempo, then, a thirty-year career could leave as little as four years in which the Army could fully leverage an officer's relevant KSAs.

The Army has several options for dealing with this implication, all of which have advantages and disadvantages. Obviously, the preferred option would be to manage promising officers very carefully in order to develop functional expertise and JIIM expertise concurrently. The increasing number of positions that offer experience in the JIIM domains may render this option feasible. If not, the Army could either seek to lengthen officer careers or accept lesser degrees of expertise in either functional or JIIM domains.

This time shortage may be even more pronounced in light of our finding that the tactical, operational, strategic, and institutional echelons require qualitatively different KSAs. The Army has long recognized the importance of developing expertise at the tactical level, as the discussion of battalion command above illustrates. The Army has also tacitly recognized the need for a distinct development path for "strategic leaders," as demonstrated by various general officer education initiatives and the establishment of functional areas in nuclear and counterproliferation matters and strategic plans and policy. Finally, the development of functional areas for force management, simulation operations, and operations research and systems analysis demonstrates a strong commitment to preparing officers to serve in the generating force.

The Army appears to devote less effort to preparing officers to serve at operational echelons, i.e., at echelons—above the tactical level—where officers can practice and gain experience in the operational level of warfare. To be sure, the Army devotes considerable effort to the study of the operational art in its various intermediate-level professional military education courses, most certainly in the School of Advanced Military Studies (SAMS), and to some extent in other venues. The missing element, however, is experience. According to JP 3-08, officers can gain operational-level experience on the staff of a combatant command or a combined joint task force. By extension, experience on a corps or theater army staff could be considered operational-level experience (FMI 3-0.1). Yet officers, especially the most successful officers, seldom spend six to ten years in these assignments.

To be sure, many competencies, especially those related to command, are probably transferable. As we observed in the last chapter, however, context is critical. Officers working at the operational level experience significantly different stimuli and seek to create significantly different effects from those at the tactical level. They have to think on a somewhat different time scale and, according to the new FM 5-0, employ significantly different thought processes (systemic operational design) from officers at tactical echelons. Functioning at operational echelons is not simply an amplified echo of service at tactical echelons.

Potential options for developing operational expertise resemble those for developing JIIM expertise. Again, the Army could simply attempt to manage officers so that their assignments provided both tactical and operational expertise more or less concurrently. As Ambassador Ron Neumann observed, commanders at the brigade level and below do, in fact, have to plan and execute campaigns over long duration.[1] It should go without saying that if brigade headquarters have to plan and execute at the operational level of war, this circumstance applies with even greater force at division headquarters. Once again, another option would be for the Army to lengthen careers. Finally, the Army could alter the balance between tactical and operational experience for a cadre of talented officers, in the manner of the Prussian General Staff. Under that system, talented officers were identified early for service on the General Staff, effectively at the operational level, and were then provided just enough experience at the tactical level to ground them in the fundamentals of their profession (Goerlitz, 1959).

Both of these issues are simply facets of the larger issues inherent in balancing breadth and depth in developing officers. This balance is never an easy one, and the U.S. Army has struck the balance differently throughout its history. Immediately after World War II, the Army prioritized breadth, an emphasis that Army leaders had thought produced the officers who had coped so successfully with the war's military and political complexities (U.S. Army, 1948; U.S. Army, Army Ground Forces, 1946). After the Vietnam War, however, policy shifted to emphasize depth somewhat more, an emphasis reflected in the concept

---

[1] Translating that experience into expertise, however, would also require the Army to thoroughly educate officers liable to serve at the brigade level in the operational art before they serve there.

of branch qualification, longer command tours, and renewed respect to the institution of command at all levels.

Both cases highlight the importance of context in striking the balance between breadth and depth. After World War II, Army leaders were preparing for another global conflict that they thought would look a lot like the war they had just finished, a war that would require officers who could collaborate well with allies and cope with the unfamiliar. After Vietnam, General William DePuy thought that the Army had to prepare for an intense, lethal, "come-as-you-are" war against the Warsaw Pact in Europe. That war would need officers who were absolute masters of the tactical, operational, and technical aspects of their profession. It is important to remember that it is highly unlikely that any balance struck will remain appropriate indefinitely. Army leaders will need to pay careful attention to the strategic context and adjust developmental patterns accordingly.

Directions for Further Research

Essentially, this study performed job analyses to identify the KSAs Army officers require in JIIM domains. Using that information, we identified the positions that conferred JIIM experience, and modeled the inventory of expertise that the Army could hope to amass in these domains under various assumptions. Further research could expand and refine this job analysis approach, focusing on the association of KSAs with various echelons. Another possible research path would focus more narrowly on the question of expertise in each domain, both describing it more precisely, and exploring in some detail how officers can develop those particular forms of expertise more effectively. Finally, further research could model the likely effects on the force of various officer-development objectives and their implementing policies.

As we noted, our sample size was not large enough to ensure that we had fully identified all JIIM knowledge, skills, and abilities, especially those associated with the intergovernmental domain. We were unable to describe in detail the manner in which KSAs differed among various functional categories, though intuition strongly suggests that they do differ. While we could ascertain that each echelon (tactical, operational, strategic, and institutional) required different sets of KSAs, we are less confident in the particular associations we identified.

Thus, an extension of these job analyses might be in order. Such an extension would seek first to identify the full range of KSAs that might be required in different domains and at different echelons. Once the full range of capabilities is identified, it should then be possible to conduct a larger-scale survey to establish a firm quantitative basis for associating particular KSAs with certain domains and echelons. In any case, continued review and analysis of the Joint Duty Assignments List is in order as more information becomes available, especially through results from the new Joint Qualification System.

Alternatively, a focused inquiry into the development of expertise might benefit the Army. Such an inquiry would seek to identify the particular kind and degree of KSAs that differentiate expert performers from the merely competent. While a variety of analytical approaches to this question exist, they all involve identifying "experts" in the profession, describing the qualities that make them expert, and then explaining the process by which they came to be experts. Such an approach could focus narrowly on expertise in the JIIM domains or, more broadly, on expertise in full-spectrum operations. The latter approach might also be able to ascertain the importance of JIIM knowledge, skills, and abilities relative to other domains of professional expertise.

Both approaches might be combined to identify the KSAs associated with expertise at the operational echelon. As discussed in the last section, Army officers spend considerably less of their organizational development time at this echelon than at the tactical, strategic, or institutional echelons. Altering that balance requires careful consideration of the nature and amount of experience appropriate to service at that level, and as well to the degree of developmental benefit such service would provide.

Finally, the Army might wish to model more precisely the effects of particular personnel policies on its ability to maintain a suitable inventory of JIIM-experienced officers. For this project we adapted a model developed for other purposes to approximate the potential for the Army to meet demand for officers with varying levels of JIIM expertise in the three major functional categories. More precise estimates would allow Army leaders to reassess the allocation of branch-immaterial positions among the branches and functional areas in order to ensure an appropriate distribution of JIIM-experienced officers at the branch and grade level of detail.

APPENDIX A

Knowledge, Skill, and Ability Definitions

Knowledge

Allied nations' capabilities, culture, and doctrine. Denotes the national capabilities (as opposed to those capabilities provided within a NATO or other framework) of the U.S. Army's habitual military partners. "Capabilities" describes the effects relevant to full-spectrum operations that other nations' forces and agencies can achieve, together with the resources and preconditions required to achieve those effects, as well as national capacity to provide those capabilities. Also implies some understanding of people of a nation and within national organizations, as needed for diplomacy and amicable interaction.

Area expertise, specific to geographical area. Denotes an understanding of the geography, society, and culture of a specific geographical area. This understanding includes awareness of the area's current political, military, economic, social, informational, and infrastructure conditions, institutions, and dynamics. Implies a broader and more textured understanding of national or regional context.

Embassy structure, processes, and culture. Understands the organization, functions, and authorities of a standard U.S. mission.

Enterprise knowledge. Enterprise knowledge consists of an understanding of how the soldier's organization fits into the parent organization and how it relates to its external environment. For example, enterprise knowledge applied to Operation Iraqi Freedom would consist of understanding that Coalition forces were conducting counterinsurgency, that the general approach was to provide population security enabling bottom-up political reconciliation, and where one's unit or headquarters fit into the overall mission.

Functional expertise. Competence within one's branch, functional area, or professional specialty (e.g., engineers). This competence is not unique to any of the JIIM domains but is essential to the successful performance of duties in a JIIM context.

General cultural understanding. Appreciating that there are significant differences between cultures and possessing working knowledge of the principal dimensions of cultural difference. Does not imply understanding of a specific culture per se but rather an understanding of what dimensions may be relevant to interpersonal interaction.

International partners' capabilities, culture, and processes. "International partners" refers to those nations and other entities, including nongovernmental organizations and international organizations like the United Nations, that are not permanently allied with the United States. "Capabilities" describes the effects relevant to full-spectrum operations that other forces and agencies can achieve, together with the resources and preconditions required to achieve those effects, as well as national capacity to provide those capabilities. Includes knowledge of an organization's culture, its employees' collective beliefs about appropriate ends for the use of capabilities, their relationship to other nongovernmental and government organizations (especially their relationship to military operations), and the appropriate means by which decisions are reached (e.g., consensus-driven or hierarchical).

Joint force capabilities and doctrine. "Joint capabilities" are differentiated from other service capabilities in that the term describes the synergies to be achieved from integration. Joint doctrine refers to capstone doctrine such as JP 1, *Unified Action,* JP 3-0, *Joint Operations,* JP 1-02, *Department of Defense Dictionary of Military and Related Terms,* and JP 5-0, *Joint Operation Planning,* a knowledge of which provides a common lexicon and analytical framework. It also includes a working knowledge of joint doctrine relevant to an officer's career field.

Joint organization and processes. Knowledge of those entities, such as defense agencies, combatant commands, and the Joint Staff, that are explicitly joint, and concomitant knowledge of their relationship to the rest of the Department of Defense. Includes an understanding of policy, the processes by which decisions are reached, and organizational responsibilities. CJCSI 3170, Joint Capabilities Integration and Development System, is an example of a joint process.

National Incident Management System. The National Incident Management System (NIMS) provides a systematic, proactive approach to guide departments and agencies at all levels of government, nongovernmental organizations, and the private sector to work seamlessly to prevent, protect against, respond to, recover from, and mitigate the effects of incidents, regardless of cause, size, location, or complexity, in order to reduce the loss of life and property and harm to the environment (FEMA, 2009).

NATO capabilities, organization, policy, doctrine, and processes. Understanding the formal structure of NATO, how it reaches political and military decisions, the processes by which it makes decisions about policy, strategy, force structure and doctrine, and the doctrine itself.

Other U.S. services capabilities, culture, and doctrine. This area of knowledge includes awareness of all other service capabilities that are relevant to the operational environment. Capabilities are the effects that other services can achieve that are relevant to full-spectrum operations; this knowledge should include the enabling conditions for these capabilities to be effective (e.g., security), and agencies' capacity to provide these capabilities. A service's culture is comprised of service members' collective beliefs about appropriate ends for the use of capabilities, their relationship to other government agencies, and the appropriate means by which decisions are reached (e.g., consensus-driven or hierarchical).

Personal acquaintance with U.S. state officials. Individual personalities play a much greater role at state level than at the federal level. Army officers' personal acquaintance with individual officials can greatly facilitate operations in the intergovernmental domain.

Stability operations (including counterinsurgency) theory and practice, beyond official joint and Army doctrine. There are other views on stability operations and counterinsurgency than those presented in FM 3-24. Capable practitioners understand those other views and the historical context in which they were developed, and are able to assess their appropriateness and apply that knowledge selectively to their individual situation.

Statutory, regulatory, and policy environment for homeland defense and civil support. Title 10, Title 32, and Title 50, among other federal laws, govern the use of military capabilities in the United States. Additionally, the Department of Homeland Security and the various states have strategy statements, policy, and guidelines on how those capabilities are to be applied. Of specific interest are those provisions governing who may direct the employment of which capabilities, and which arm of government pays for which actions. Includes an understanding of the National Guard's structure and capabilities relevant to homeland defense and civil support.

Strategic issues. Currency with the major strategic problems confronting the United States.

U.S. Army capabilities, culture, and doctrine. Comprehensive understanding of the complete range of capabilities that the Army could provide that are relevant to the operational context. Such understanding may well exceed the range of capabilities within an officer's functional category. Army officers must be able to understand that the U.S. Army has its own distinctive culture, and be able to objectively analyze the differences and similarities with other organizational and national cultures.

U.S. Army organization and processes. Understanding of how the Department of the Army is organized, to include not only the Operational Army but also the Generating Force. Includes understanding how the Department of the Army plans, programs, and budgets, and the relationship of the Department of the Army to other services, the Department of Defense, and combatant commanders.

U.S. government agencies' capabilities, culture, and processes. Capabilities are the effects that other agencies can achieve that are relevant to full-spectrum operations; this knowledge should include the enabling conditions for these capabilities to be effective (e.g., security), and agencies' capacity to provide these capabilities. An agency's culture consists of its employees' collective beliefs about appropriate ends for the use of capabilities, their relationship to other government agencies, and the appropriate means by which decisions are reached (e.g., consensus-driven or hierarchical). Processes are the formal means by which decisions are made and resources allocated.

U.S. government budget policy and processes. Understanding how the U.S. government authorizes and appropriates monies relevant to full-spectrum operations, the processes and timelines by which that money is obligated, and the limitations on who can spend it and how.

U.S. government strategy and policy. A thorough grasp of the U.S. government's current and historical declaratory policy and strategy, further informed by actual practice. This knowledge should include not only activities in the international arena, but also policy with regard to supporting institutions.

U.S. state governments' organization and structure. Each of the 54 states and territories have unique, albeit similar, constitutional arrangements that establish the authority and responsibilities of various state officials and the functions of executive departments. Further, many have different jurisdictional arrangements that apportion authority and responsibility differently among towns/cities, counties/parishes, and state government. Finally, there are often prominent nongovernmental organizations (e.g., the Farm Bureau) that can affect the conduct of support operations or lend informal support to expeditionary operations.

Skills

Active learning. Understanding the implications of new information for both current and future problem solving and decisionmaking (O*NET, 2003).

Coordinating across national cultures. Adjusting U.S. Army, joint, and interagency organizations' actions in relation to those of other organizations representing a different national culture, where national culture constitutes the principal point of divergence. Coordination between U.S. Army elements and army organizations from another nation is an example of this skill.

Coordinating across organizational cultures. Adjusting U.S. Army organizations' actions in relation to those of other U.S. organizations, usually nonmilitary, where organizational culture constitutes the principal point of divergence. Coordination between U.S. Army elements and other U.S. government agencies is an example of this skill.

Critical thinking/judgment and decisionmaking. Considering the relative costs and benefits of potential actions to choose the most appropriate one. Analysis of complex problems to determine appropriate solutions (O*NET, 2003).

Cross-cultural communication skills. Skill at application of general communication skills with a particular emphasis on conveying and receiving meaning and intent across organizational and national boundaries.

Employing joint capabilities. While one must understand joint capabilities in order to employ them, employment speaks to the actual skill with which they are integrated into operations. It includes obtaining and allocating joint capabilities and the degree to which the capability employed is capable of attaining its desired effect.

General communication skills. Communicating effectively in writing or orally as appropriate for the needs of the audience; understanding written or oral communications from others; and giving full attention to what other people are saying, taking time to understand the points being made, and asking questions as appropriate in order to aid comprehension (O*NET, 2003).

Joint Operations Planning Process. This skill refers not only to planning skills but also to the ability to manage the formal processes in order to secure the required approval and support. The Joint Operations Planning Process is an orderly, analytical process that consists of a logical set of steps to analyze a mission; develop, analyze, and compare alternative courses of action against criteria of success and each other; select the best course of action; and produce a joint operation plan or order. Also called JOPP. See also joint operation planning (JP 5-0).

Management of financial resources. Determining how money will be spent to get the work done, and accounting for these expenditures (O*NET, 2003).

Management of personnel resources. Motivating, developing, and directing people as they work, identifying the best people for the job[1] (O*NET, 2003).

Military Decision Making Process (MDMP)/Planning. MDMP is a planning tool that establishes procedures for analyzing a mission; developing, analyzing, and comparing courses of action against criteria of success and against one another; selecting the optimum course of action; and producing a plan or order. A successful planner can adapt the MDMP for application in JIIM contexts.

Negotiation/conflict resolution/persuading/influencing. Bring others together and try to reconcile differences; persuade others to change their minds or behavior (O*NET, 2003).

People skills/building and maintaining relationships. In the context of our investigation, people skills are those that allow an individual to foster positive interaction with his or her counterparts and co-workers; the ability to build and maintain relationships appears to be a major aspect of people skills. Relationship building is a technique in which

---

[1] This O*NET definition corresponds closely with the military definition of leadership. In the context of our study, the definition's most important component was "identifying the best people for the job."

practitioners build positive rapport and a relationship of mutual trust, making counterparts more willing to support requests. Examples include showing personal interest in an individual's well-being, offering praise, and understanding a counterpart's perspective (social perceptiveness). This technique is best used over time. It is unrealistic to expect it can be applied hastily when it has not been previously used. With time, this approach can be a consistently effective way to gain commitment from other individuals (adapted from FM 6-22, para 7-17).

Project management, including change management. "The discipline of planning, organizing, and managing resources to bring about the successful completion of specific project goals and objectives. A project is a finite endeavor (having specific start and completion dates) undertaken to create a unique product or service which brings about beneficial change or added value" (Wikipedia).

Training management. The process used by Army leaders to identify training requirements and subsequently plan, resource, execute, and evaluate training. Officers should be able to apply the principles of Army training management outside of a U.S. Army context.

Abilities

Adaptability/flexibility. Adaptability refers to the degree to which adjustments are possible in practices, processes, or structures of systems to projected or actual changes in conditions or circumstances. Adaptation can be spontaneous or planned, and be carried out in response to or in anticipation of changes. Many respondents referred to this in terms of an ability to tolerate ambiguity, defined as the ability to function effectively without fully understanding all the factors relevant to a given context and without clearly and explicitly defined objectives.

Deductive and inductive reasoning. The ability to apply general rules to specific programs to produce answers that make sense and the ability to combine pieces of information to form general rules or conclusions (O*NET, 2003).

Originality. The ability to come up with unusual or clever ideas about a given topic or situation or to develop creative ways to solve a problem (O*NET, 2003).

Problem sensitivity. The ability to tell when something is wrong. Does not involve solving problems, but can certainly be a short cut in problem identification and specification (O*NET, 2003).

Self-awareness. Being objectively aware of one's traits, feelings, and behaviors (O*NET, 2003).

APPENDIX B

Interview and Focus Group Protocol

Introduction

I'm ______________ from the RAND Corporation and this is my colleague, ________________. We're working on a study for the U.S. Army Human Resources Command to identify and describe the knowledge, skills and abilities associated with increasing levels of professional maturity in the joint, interagency, intergovernmental and multinational domains. (If the subject exhibits any confusion at all about these terms, refer to Enclosure A, Definitions.) We're trying to identify the KSAs needed for individuals who work in a joint, interagency, intergovernmental or multinational context, and what the Army needs to do to help them develop those KSAs.

We'd like to discuss your most recent (joint, interagency, intergovernmental, and/or multinational) assignment. If you'd prefer, we could also discuss your most memorable job.

Before we begin, however, we need to go through a few necessary formalities to comply with U.S. government and RAND Corporation policies regarding Human Subjects Protection. *[The research team complied thoroughly with all such policies throughout the course of the project.]*

Role Analysis

1. Let's start with your job. Where did you work, and what did you do there?

2. Could you briefly describe your role in that organization? What were your major tasks?
3. Here's a list of potential tasks for your job. Which of these tasks did you do? Which did you not do? How accurate is the list, anyway? *[Discontinued after the first several interviews.]*
4. Based on your experience, would you say that some of these tasks are particularly important for your job?
5. For the tasks you have indicated were important, what was particularly (joint, interagency, intergovernmental or multinational) about each?
6. Did you spend a lot of time on these tasks? Was there anything that wasn't important per se, but took up a lot of time?

[Repeat as needed]

KSA Decomposition

Now we'll move on to discuss knowledge, skills and abilities relevant to the tasks you've identified. We'll begin by describing what we mean by those terms.

- **Knowledge:** Organized sets of principles and facts applied in general work settings. Examples: Joint doctrine, service doctrine, national strategic documents.
- **Skills:** Developed capacities to perform tasks, predicated in part on the individual's possession of relevant underlying abilities. In a civilian context, skills include things like writing, reading comprehension, etc. Skills can also include things like managing a certain process, e.g., a Joint Staff Action Package, or a type of analysis unique to one's job.
- **Abilities:** Enduring attributes of the individuals that influence performance, regarded as traits in that they exhibit some degree of stability over time. Examples include fluency of ideas, originality, the ability to work under pressure, etc.

7. To what extent did the tasks you've described require similar sets of knowledge, skills and abilities?
8. When you were performing these task(s), what bodies of knowledge did you find useful? For example, which reference(s) did you use most frequently?
9. Is there anything you wish you would have known before you took this job?
10. When performing these tasks, which skills did you use most frequently? Please describe both the general skills you employed, such as writing or oral presentation skills, and those which were specific to your job's (joint, interagency, intergovernmental or multinational) aspects?
11. Are there any skills that you wish you had acquired before you took this job?
12. When performing these tasks, which abilities proved most relevant? Remember, abilities are relatively enduring personal traits. Some may be more salient in a particular context than in others, but a person's abilities do not change quickly over time.

(Repeat as necessary for each grouping of tasks)

Developmental History

13. Which developmental experiences did you find the most helpful in preparing you for these jobs?
14. In what ways did that/those experience(s) prepare you—for example, did you develop a skill in that prior experience, or did you learn some facts (acquire some knowledge) that you need to apply to the task in your current or most recent job?
15. (If no): Have you had any experiences since that you feel would have helped you prepare to perform your tasks? Why do you say you weren't prepared—is there some skill you needed to develop with OJT, were there some facts or information you ideally would have had before doing the task in your current job?

16. Can you tell me more about that prior experience? Was there some specific training or a class you had that prepared you to do task X?
17. If yes: what did you learn, or what did you learn how to do? Did you learn any additional skills, knowledge, or further develop any abilities in that developmental experience?
18. So if I understand you correctly, it sounds like these knowledge, skills, and abilities are important for doing this task. [Summarize what respondent has said.] Does that sound right? Is there anything else you'd add to the list?
19. VITAL: Before we leave, is there anyone else to whom you would recommend we talk? For instance, do you recall colleagues who were particularly good at their JIIM job, or supervisors who might provide us with a useful perspective on what was required?

Enclosure A: Definitions

Joint: Activities involving two or more military services in pursuit of a common end.

Interagency: Activities involving two or more U.S. government agencies, including the Department of Defense, to achieve a common end.

Intergovernmental: KSAs that enable soldiers to facilitate the attainment of synergies between the Department of Defense and local, regional and state authorities.

Multinational: Activities which involve U.S. Department of Defense organizations with the military forces of other nations under the rubric of a contingent alliance or coalition.

Enclosure B: Task List

Tasks were abstracted from a number of sources, including but not limited to National Security Presidential Directive 44 (NSPD-44), the Department of State's Coordinator for Reconstruction and Stabilization's Essential Tasks Matrix, Department of Defense Directive 3000.05, and the Universal Joint Task List.

Institutional Tasks

Acquisition

Assess joint capabilities (warfighting analysis)

Conduct force development

Coordinate Department of Defense budget and program submissions with other government agencies

Coordinate draft Department of Defense policy with services, combatant commands, and other defense agencies

Coordinate funding of U.S. government stabilization and reconstruction activities

Coordinate mapping, charting and geodesy

Coordinate mobilization

Coordinate the development of guidance on the global employment of the force (GEF)

Develop guiding precepts, policy and implementation procedures for integrated U.S. government stabilization and reconstruction operations

Develop recommendations for establishing executive branch management and oversight mechanisms for specific contingency operations

Develop strategies to build partnership security capacity abroad and to seek to maximize nongovernmental and international resources for reconstruction and stabilization

Develop widely applicable scenarios for use in force development supporting integrated U.S. stabilization and reconstruction operations

Develop/assess joint doctrine

Develop/assess joint policy

Identify and track issues requiring analysis in the context of the development of the defense program

Identify information sources and contacts in countries and regions in which the U.S. may have to conduct stabilization and reconstruction activities

Identify lessons learned and incorporate them into operations

Industrial management

Joint program management

Lead U.S. government development of a strong civilian response capability

Participate in the development of defense planning scenarios

Participate in the development of the strategic planning guidance

Participate in the development of the defense strategy

Provide legal advice and analysis

Research, development, testing, evaluation and simulations

Resolve relevant policy, program and funding disputes among U.S. government departments and agencies with respect to U.S. foreign assistance and foreign economic cooperation

Resource/financial management

Support global force management

Strategic Tasks

Assess strategic effects

Conduct strategic level intelligence analysis

Coordinate and integrate U.S. stabilization and reconstruction efforts and preventative efforts with other U.S. government agencies, intergovernmental organizations, non-governmental organizations

Coordinate counterproliferation

Coordinate development of warplans

Coordinate for the establishment of indigenous and international mechanisms for prosecuting war crimes and crimes against humanity

Coordinate funding of U.S. government stabilization and reconstruction activities

Coordinate with indigenous and international security forces who are parties to the conflict

Counter or manage deterrence of CBRNE weapons

Develop and assess U.S. global force posture

Develop detailed contingency plans for integrated U.S. government stabilization and reconstruction efforts

Develop strategies for foreign assistance and economic cooperation directed towards states and regions at risk of or emerging from conflict

Develop, conduct or provide intelligence, surveillance and reconnaissance at a strategic level

Develop, coordinate, and/or revise country strategic plan/campaign plan

Develop, coordinate, or assess theater security cooperation plan

Foster interagency relations

Foster intergovernmental relations

Foster multinational relations

Integrate military strategic plan into embassy mission performance plan

Plan and conduct deployment, redeployment of forces

Plan and coordinate global force protection

Plan for and coordinate humanitarian de-mining for a given country or region

Plan for and coordinate measures to ensure food security for contested population

Plan for and coordinate the disposition of refugees and internally displaced persons

Plan for and coordinate the establishment of an effective corrections system that meets at least minimum acceptable international standards

Plan for and coordinate the establishment of an effective host nation justice system

Plan for and coordinate the establishment of host nation police forces

Plan for and coordinate the establishment of public order and safety

Plan for and coordinate the establishment of territorial security

Plan for and coordinate the provision of shelter and non-food emergency relief

Plan for the disposition of contending armed forces, paramilitary organizations, intelligence services and belligerents

Plan for the protection of indigenous individuals, infrastructure and institutions for a given country or region

Provide strategic direction and integration

Resource/financial management

When necessary, identify appropriate issues for resolution or action through the National Security Council interagency process

Operational Tasks

Conduct civil-military operations

Conduct operational net assessment

Deploy and redeploy forces from theater

Employ joint fires

Establish theater force requirements and readiness

Mapping, charting, and geodesy

Neutralize enemy information systems

Plan for and employ joint air and missile defense

Plan for, coordinate, and conduct special operations

Plan, conduct, and coordinate operational maneuver

Plan, coordinate, and assist in host nation basing

Plan, coordinate, and execute operational ISR

Plan, coordinate, and execute theater basing

Plan, coordinate, and execute theater health service support

Protect friendly information systems

Provide operational legal support

Provide or coordinate protection of the force, or protect the force

Provide or exercise operational (level of war) command and control

Provide security to host nation populations

Provide operational sustainment support

Resource/financial management

Sustain theater forces communications systems

Tactical Tasks

Employ joint fires capabilities

Employ joint battle command capabilities

Employ joint intelligence, surveillance and reconnaissance capabilities

Employ joint sustainment capabilities

Employ joint force projection capabilities

Train indigenous forces in maneuver warfighting function

Train indigenous forces in fires warfighting function

Train indigenous forces in intelligence warfighting function

Train indigenous forces in command and control warfighting function

Train indigenous forces in sustainment warfighting function

Train indigenous forces in protection warfighting function

Coordinate civil works projects

Support local governance by providing security

Support local governance by providing advice on administrative and governing functions

Support local governance by providing sustainment support

Provide humanitarian assistance to affected population

Coordinate provision of humanitarian assistance with other U.S. government agencies

Coordinate the provision of humanitarian assistance with local, state and federal authorities

Provide security to indigenous population

Analyze PMESII dynamics in an area of operations

Develop detailed contingency plans for integrated stability and reconstruction operations for at risk states and regions

Develop detailed contingency plans for integrated stability and reconstruction operations for at risk states and regions

Separate contending parties

Disarm and demobilize former combatants

Substitute for or supplement local police

Mentor indigenous military forces

Establish secure environment

Acquire information on potential security threats

Acquire information on organizational structure of local security forces

Acquire information on sociopolitical conditions within country

Obtain information on organization of local military and police forces

Train foreign forces

Conduct basic law enforcement operations

Conduct counter narcotics operations

Conduct counter insurgency operations

Secure elections

Detain people suspected of criminal or unlawful actions

Use interpreters and military police officers

Capture war criminals

Conduct cordon and search operations

Conduct strikes and raids against terrorists and their infrustructure

Train indigenous personnel in crisis management, airport security, terrorist financing, and border control

Work with international and local agencies during elections to help ensure integrity

Conduct weapons survey

Collect weapons

Arrange weapons storage, reutilization or destruction

Disband formerly armed unit and reduce number of combatants in an armed group

Register, count and monitor combatants

Integrate former combatants into new or reformed local security units

Track changes in behavior of population

Gauge potential reactions of population to operation

Participate in civic action or small-scale reconstruction programs

Provide security shield for early humanitarian relief and reconstruction activities

Provide health care

Supply water and food to the population

Deal with mines and hazardous materials

Institute quarantine measures in the event of communicable disease outbreaks

Perform emergency repairs of infrastructure

Provide retraining and new equipment to individuals who successfully pass vetting process

Train indigenous forces in skills necessary for effective operations and maintenance of equipment

Enclosure C: Lists of Knowledge, Skills and Abilities

The lists of knowledge, skills and abilities found on O*NET served as our starting point, supplemented by the research team's analysis of the domain-specific knowledge, skills and abilities implied by the tasks listed in Enclosure B.

Knowledge Area

Joint doctrine

Army doctrine

NATO doctrine

Department of Defense Planning, Programming and Budgeting System

Joint Capabilities Integration and Development System processes

Current USG policy and strategy

Area expertise (history, geography, culture)

Capabilities and limitations of other governments' military forces

Infrastructure repair, development and maintenance

Capabilities, limitations and culture of other U.S. government agencies

DARPA structure, processes and culture

U.S. Air Force structure, processes and culture

U.S. Navy structure, processes and culture

U.S. Marine Corps structure, processes and culture

U.S. Central Command structure, processes and culture

Joint Staff structure, processes and culture

Office of the Secretary of Defense structure, processes and culture

Defense Intelligence Agency

Drug Enforcement Agency structure, processes and culture

USAID structure, processes and culture

EPA structure, processes and culture

National Geospatial Agency structure, processes and culture

National Reconnaissance Office structure, processes and culture

NSA structure, processes and culture

Office of Management and Budget structure, processes and culture

JFCOM structure, processes and culture

TRANSCOM structure, processes and culture

Capabilities, limitations and culture of non-governmental organizations and international organizations

Foreign assistance

National security council structure, processes and culture

U.S. government budget process

Theater logistics and services

Command, control, communications and computers

Civil affairs and psychological operations

Special operations

Counterinsurgency theory and practice (beyond U.S. doctrine)

Counterterrorism theory and practice (beyond U.S. doctrine)

Information operations (beyond U.S. doctrine)

Space operations (beyond U.S. doctrine)

Skills

Joint operations planning processes

Campaign planning

Force management

Training management

Establishing effective organizational structures

Developing and coordinating a Joint Staff Action Package

Managing a study

Managing organization's strategic communications

Request for forces/request for capabilities process

Written communication

Active listening

Oral communication

Critical thinking

Active learning

Monitoring [performance]

Social perceptiveness

Negotiation

Instructing

Complex problem solving

Operations analysis

Systems analysis

Judgment and decision making

Management of financial resources

Management of personnel resources

Abilities

Implementing plans and directives

Functioning as a change agent

Coordinating work across external divisions and organizations

Developing and maintaining good relations with subordinates

Developing and maintaining good relations with stakeholders

Self-awareness

Adaptability

Flexibility

Ability to deal with ambiguity

Long-term visionary outlook

Breadth

Ability to translate complexities into clear direction

Analytical perspective

Abstract problem solving

Attention, concentration and vigilance

People perspective

Oral comprehension

Fluency of ideas

Originality

Problem sensitivity

Deductive reasoning

Inductive reasoning

Information ordering

Flexibility

Memorization

Speed of closure (The ability to quickly make sense of, combine and organize information into meaningful patterns)

Flexibility of closure (The ability to identify or detect a known pattern that is hidden in other distracting material)

Perceptual speed

Visualization

APPENDIX C

Survey Instructions

MEMORANDUM FOR

Assignments Officers, Army Human Resources Command
Assignments Officers, Army Senior Leader Development Office

SUBJECT: Completion of survey on joint, interagency, intergovernmental and multinational knowledge, skills and abilities

1. We request that you forward this memorandum to all assignments officers in your branch, and any other individuals that you think could contribute to this survey.

2. The purpose of this survey is to assess the degree to which the kinds of positions you manage strongly develop knowledge, skills and abilities that contribute to one or more of the joint, interagency, intergovernmental or multinational domains. By "strongly develop," we mean that demonstration of a high degree of proficiency in the knowledge, skill or ability in question is either essential to receiving an "above center mass" rating in that position, or critical to success. By "critical," we mean that failure to demonstrate the knowledge, skill or ability in question will result in either relief for cause or immediate transfer to another duty position.

3. Participation in this survey is entirely voluntary, though the results of your input are extremely important to the Officer Personnel Management System Task Force and the Army Human Resources Command. All your responses will remain in confidence, known only

to you and the RAND Corporation research team. For more on your rights as a respondent, please read Encl 1.

4. Instructions for completing the survey form are as follows:
   a. The survey form is a Microsoft Excel-based workbook with embedded macros.
   b. On receiving an electronic copy of the file ("JIIM Assignments Officer Survey.xls"), save the file under a new name according to the following protocol: the branch or functional area you manage, followed by the highest rank you manage, e.g., "INFCPT.xls." NOTE: The important thing is that the file get a different name, to allow subsequent tracking and merging.
   c. When you open the file, you will view a dialog box asking whether you wish to enable macros. Click "Enable Macros."

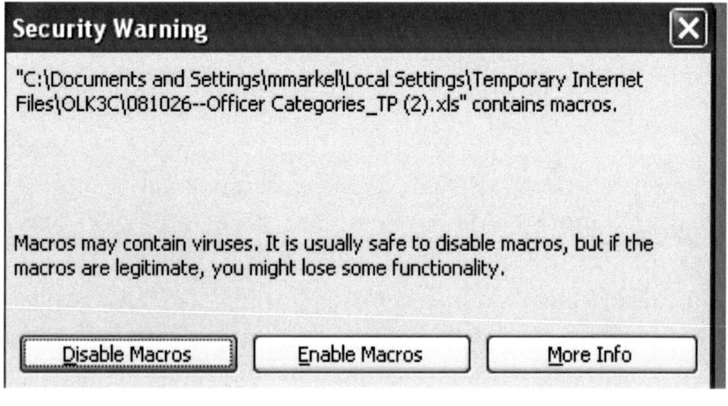

RAND *MG990-C.1*

   d. Go to the worksheet on the tab "First Page." You will answer a series of questions as depicted below, most of which are self explanatory. When you indicate your branch/functional area, it will change the hyperlink at the bottom of the page. Once you answer the other questions, click the "Update" button, which is almost at the bottom of the page. After clicking on the "Update" button, click the hyperlink listing your branch/functional area at the bottom to take you to the sheet listing positions in your branch/functional area. Answer the questions there, which will provide a basis for describing the collective expertise applied to this effort. After completing the "First Page," use the hyperlink for your branch at the bottom right hand corner of the page to jump to your worksheet in the questionnaire. Each branch, functional area or area of concentration has its own worksheet.

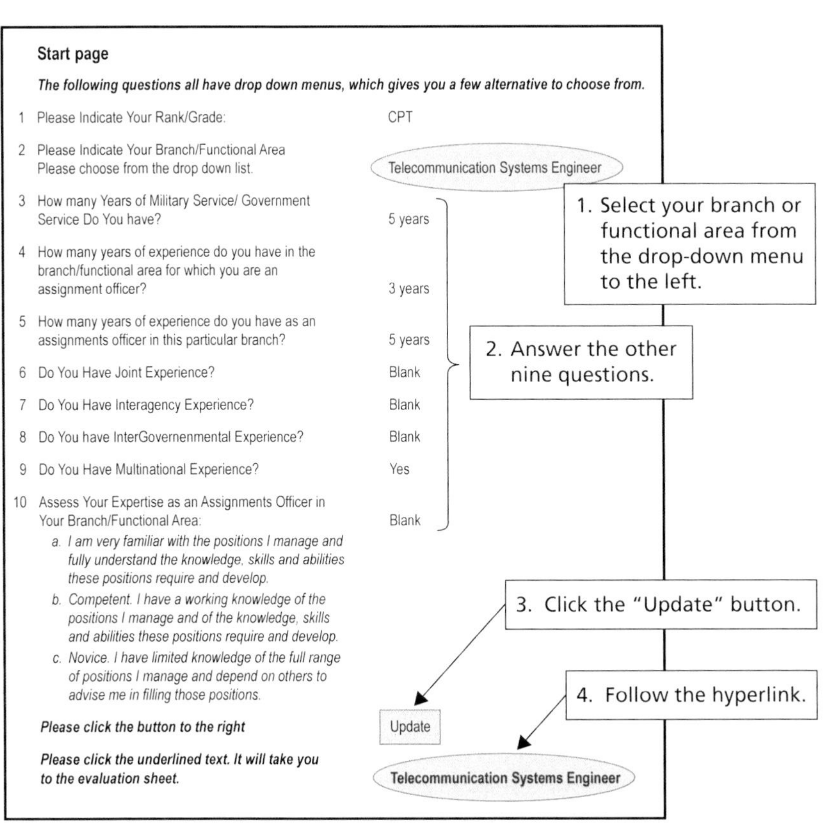

RAND *MG990-C.2*

e. The hyperlink will take you to the worksheet for your branch or functional area, which lists all the kinds of position that exist within your branch/career field according to DA PAM 600-3. The sheet lists the kinds of position within the branch in column "B," the knowledge, skills and abilities in Row "1," and has a remarks column in column C. The panes are "frozen" so you can keep the position in view as you scroll horizontally across the knowledge, skills and abilities columns. If for some reason the hyperlink does not work, you can find the worksheet for your branch/functional area by scrolling through the tabs at the bottom of the page.

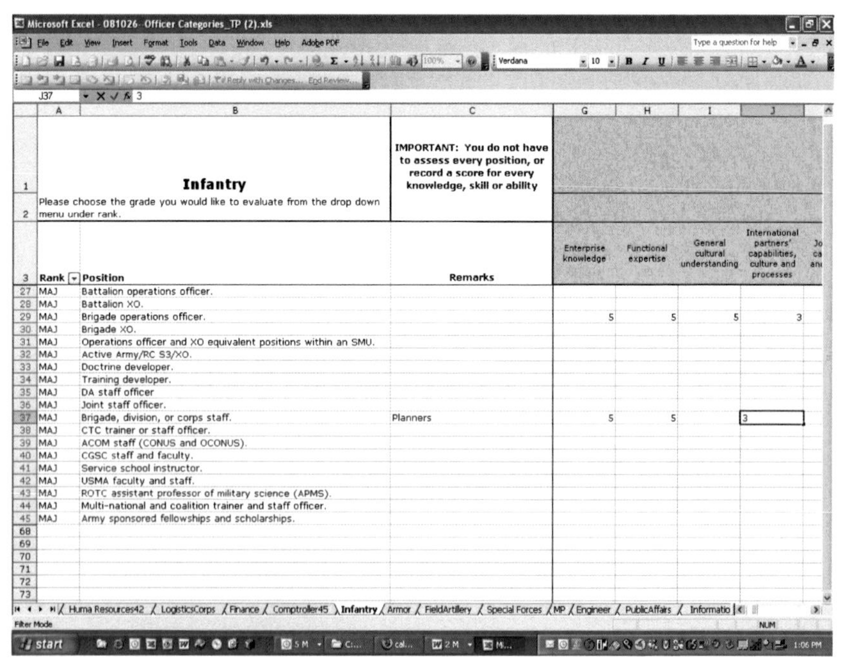

RAND *MG990-C.3*

f. If you choose, you can filter for the rank(s) you manage using the drop down menu in Column B, "Rank." If you manage several ranks, you may wish to assess one rank at a time, e.g., all lieutenants, followed by all captains, etc. Alternatively, you can simply scroll through the various ranks to find those which you manage.

g. For those positions that strongly develop several of the knowledge, skills and abilities listed and defined at Encl 2, assess the degree to which that position develops the knowledge, skill or ability in question on a scale of one to five, with five ("5") being the strongest. You will have to type in the number, as indicated in the figure below. There are over forty knowledge, skills and abilities, so you will have to scroll laterally through the columns. You do not have to assess every position, nor do you have to assess every KSA for a given position. If you have any question at all, you should leave the cell blank.

h. You should save your work periodically. When you are done, save your work again.

i. Email the file directly to either mmarkel@rand.org, csims@rand.org, or tpanis@rand.org. To preserve your anonymity, you should not send the file to your supervisor or any other person within Army Human Resources Command. Upon receiving your email, the survey team will save the survey form and delete your email in order to preserve your anonymity.

6. If you have any questions concerning this survey, please contact

M. Wade Markel
RAND Corporation
703.413.1100 x 5108
mmarkel@rand.org

Encl 1, Informed Consent

Informed Consent, Assignment Officer Survey

This survey is part of a study conducted by the Arroyo Center of the RAND Corporation in Santa Monica, California and sponsored by the Human Resources Command and Army G-1. The goals of the study are to describe the knowledge, skills and abilities (KSAs) associated with joint, interagency, intergovernmental and multinational (JIIM) domains for successive stages of officer careers, and to develop a framework that will help the Army to better track and manage these skills in its officer inventory.

Participation in this survey is voluntary. You are also are free to skip any question or portion you would prefer not to answer. You will not be required to provide any information about your identity. The estimated time to complete the questionnaire is about 15 to 20 minutes.

Your answers on this questionnaire will go directly to the research team at the RAND Corporation. Your responses will be anonymous. While we will ask you to send us the form directly by email, we will save the file and delete your email soon after receipt. If you choose to contact our team and provide your contact information, RAND will keep this information confidential and maintain it only as long as necessary to determine whether follow-up is needed. After successful follow-up or at the conclusion of the study, all contact information will be destroyed.

Your participation is very important to the study team's efforts to get as complete a picture as possible of the knowledge, skills and abilities associated with various U.S. Army experiences.

If you have any questions about the study or your participation, you may contact the RAND project leader, Wade Markel, at (703) 413-1100, ext. 5108.

If you have any questions about your rights as a research participant, you may contact:

Jim Tebow, Co-Administrator
RAND Human Subjects Protection Committee
1776 Main Street M3W
Santa Monica, CA 90407-2138
(310) 393-0411 x7173
James_Tebow@rand.org

Bibliography

Project Interviews and Focus Groups

al-Sondani, Brigadier General Hussain, Defense Attaché, Embassy of Iraq, interview with Wade Markel, Washington, D.C., August 6, 2008.

Anderson, Brigadier General Stephen, former Deputy Chief of Staff for Resources and Sustainment, Multi-National Force–Iraq, interview with Wade Markel, May 12, 2008.

Aylward, Brigadier General Pete, Director, Anti-Terrorism and Homeland Defense, Joint Staff J-3, interview with Maren Leed, Washington, D.C., August 8, 2008.

Bame, David, Director, Office of Regional Security and Arms Transfers, Bureau of Political-Military Affairs, U.S. Department of State, interview with Carra Sims and Chip Leonard, Washington, D.C., June 4, 2008.

Billings, Captain Rob, Chief of Plans, State of Louisiana, telephone interview with Wade Markel, December 5, 2008.

Boden, Lieutenant Colonel Mike, interview with Major Rob Thornton, April 2, 2008.

Carrigg, Lieutenant Colonel Kelly Marie, U.S. Army Training and Doctrine Command Liaison to the French Army's Center for Doctrine for Force Employment, telephone interview with Wade Markel, May 1, 2008.

Cheek, Brigadier General Gary, former commander, Regional Command–East, Combined Joint Task Force–76, interview with Pete Schirmer, Arlington, Va., July 15, 2008.

Christensen, Colonel Jay, former Director, Sustainment, Multi-National Corps–Iraq (MNC-I), interview with Wade Markel, Arlington, Va., June 4, 2008.

Curry, Colonel Pete, former Director of Operations, International Security Assistance Force, telephone interview with Jim Crowley, October 1, 2008.

Dailey, Ambassador Dell, Coordinator for Counterterrorism, U.S. Department of State, interview with Maren Leed, Washington, D.C., August 7, 2008.

Dobbins, Ambassador (ret.) Jim, former U.S. Ambassador to Afghanistan, interview with Wade Markel and Chip Leonard, Arlington, Va., March 10, 2008.

Dolan, Colonel (ret.) Bill, former Director of Studies and Analysis, Joint Center for Operational Analysis, U.S. Joint Forces Command, interview with Maren Leed, Washington, D.C., July 10, 2008.

Dornblaser, Colonel David, Director of Plans, Policies and Weapons, U.S. Army Security Assistance Command, interview with Carra Sims and Pete Schirmer, April 24, 2008.

Fastabend, Major General David, former Director of Operations, Multi-National Force–Iraq, interview with Chip Leonard and Wade Markel, Washington, D.C., April 29, 2008.

Fastabend, Major General David, former Commander, Northwestern Division, U.S. Army Corps of Engineers, interview with Maren Leed, Washington, D.C., July 11, 2008.

Freier, Lieutenant Colonel Nathan, former Assistant for Strategy and Force Planning, Office of the Under Secretary of Defense (Policy), interview with Wade Markel, Washington, D.C., 2008.

Geehan, Major General (ret.) Brian, former Director, J-4, U.S. Central Command, interview with Chip Leonard, Arlington, Va., September 10, 2008.

Goss, Lieutenant Colonel Tom, former action officer, NATO International Military Staff, telephone interviews with Wade Markel, August 7 and 11, 2008.

Grimes, Commander (U.S. Navy) James, former executive officer to the Deputy Chief of Staff for Resources and Sustainment, Multi-National Force–Iraq, interview with Wade Markel, May 30, 2008.

Harriman, Colonel (ret.) Anthony, Senior Director for Afghanistan, National Security Council, interview with Wade Markel and Carra Sims, Washington, D.C., February 29, 2008.

Hoehn, Andy, former Deputy Assistant Secretary of Defense for Strategy, interview with Wade Markel, Arlington, Va., March 14, 2008.

Hoffman, Colonel (ret.) Hugh, Director of Security Cooperation, Office of the Secretary of Defense, interview with Wade Markel, Washington, D.C., April 9, 2008.

Jones, Brigadier (UK Army) Phil, former Director, Strategic Plans and Policy (CJ-5), U.S. Combined Joint Task Force-180, interview with Wade Markel, Washington, D.C., August 15, 2008.

Kearney, Lieutenant General Frank, Deputy Commander, U.S. Special Operations Command, interview with Maren Leed, Washington, D.C., July 17, 2008.

Kingett, Captain (U.S. Coast Guard) Frank and Lieutenant Colonel (USMC) Patrick Gramuglia, interview with Wade Markel, Arlington, Va., June 6, 2008.

Lamm, Colonel (ret.) David, former Chief of Staff, Combined Forces Command–Afghanistan, telephone interview with Wade Markel, April 10, 2008.

Lee, Colonel Fitz, and Lieutenant Colonel Dave Keefe, Department of Defense and Department of the Army Liaisons to the Office of the Coordinator for Reconstruction and Stabilization, U.S. Department of State, interview with Carra Sims and Wade Markel, Washington, D.C., May 28, 2008.

Lynes, Colonel (USMC, ret.) Jerry, Chief, Joint Education and Doctrine Division, U.S. Joint Staff J-7, interview with Pete Schirmer and Maren Leed, Washington, D.C., April 23, 2008.

MacFarland, Colonel (Promotable) Sean, former commander, 1st Brigade, 1st Armored Division, interview with Pete Schirmer and Wade Markel, Washington, D.C., March 31, 2008.

McGraw, Colonel Donald, former Director of Operations, Combined Forces Command–Afghanistan, interview with Pete Schirmer and Maren Leed, Washington, D.C., May 20, 2008.

McLauchlin, Matthew, Senior Advisor to the U.S. Ambassador, Kabul, interview with Wade Markel, Arlington, Va., April 29, 2008.

Morrison, Colonel David "Mo," former Director for Intelligence (CJ-2), Combined Joint Task Force-180, interview with Wade Markel, Washington, D.C., May 21, 2008.

Neumann, Ambassador (ret.) Ron, former U.S. Ambassador to Afghanistan, Algeria, and Bahrain, interview with Wade Markel, Washington, D.C., July 3, 2008.

Parker, Michelle, former Development Advisor to Commander, International Security Assistance Force, interview with Wade Markel, April 8, 2008.

Quantock, Colonel Mark, former commander, 205th Military Intelligence Brigade, supporting Multi-National Corps–Iraq, interview with Wade Markel, June 13, 2008.

Rattan, Commander (U.S. Navy Reserve) Patrick, former action officer, Directorate of Resources and Sustainment, Multi-National Force–Iraq, telephone interview with Wade Markel and Jim Crowley, August 21, 2008.

Richardson, Rear Admiral John, Director, Strategic Plans and Policy, U.S. Joint Forces Command, interview with Maren Leed, Washington, D.C., July 11, 2008.

Ruf, Colonel Jim, former Commander, Jalalabad Provincial Reconstruction Team, interview with Pete Schirmer, Carlisle Barracks, Pa., May 5, 2008.

Ryan, Brigadier General Mike, Director of Operations, Allied Rapid Reaction Corps, telephone interview with Wade Markel, August 18, 2008.

Stoneman, Major Mark, former battery commander, Diyala, Iraq, interview with Carra Sims and Wade Markel, March 26, 2008.

Toczek, Lieutenant Colonel Dave, Branch Chief, Plans, U.S. Strategic Command, telephone interviews with Wade Markel, June 18 and 23, 2008.

Vaughn, Lieutenant General Clyde, Director, U.S. Army National Guard, interview with Wade Markel and Chip Leonard, October 31, 2008.

Zabel, Captain Mike (U.S. Navy), former Chief, Mobility, Distribution and Logistics, Multi-National Force–Iraq, interview with Wade Markel, May 30, 2008.

U.S. Army Center for Army Analysis Analysts, focus group with Wade Markel and Chip Leonard, Fort Belvoir, Va., April 4, 2008.

U.S. Army Command and General Staff School Students, Maneuver, Fires, and Effects functional category, focus group with Wade Markel, Fort Leavenworth, Kan., July 29, 2008.

U.S. Army Command and General Staff School Students, Operations Support functional category, focus group with Carra Sims, Fort Leavenworth, Kan., July 29, 2008.

U.S. Army Command and General Staff School Students, Force Sustainment functional category, focus group with Carra Sims and Wade Markel, Fort Leavenworth, Kan., July 30, 2008.

U.S. Army War College Students, focus group with Wade Markel, Pete Schirmer, and Chip Leonard, Carlisle Barracks, Pa., 11:45 AM, May 5, 2008.

U.S. Army War College Students, focus group with Wade Markel, Pete Schirmer, and Chip Leonard, Carlisle Barracks, Pa., 2:00 PM, May 5, 2008.

U.S. Army War College Students, focus group with Wade Markel and Pete Schirmer, Carlisle Barracks, Pa., 2:00 PM, May 6, 2008.

U.S. Army War College Students, focus group with Pete Schirmer and Chip Leonard, Carlisle Barracks, Pa., 11:45 AM, May 6, 2008.

U.S. Department of State, Bureau of Political Military Affairs, focus group with Chip Leonard and Carra Sims, Washington, D.C., June 4, 2008.

U.S. Department of State, Office of the Coordinator for Reconstruction and Stabilization, focus group with Carra Sims, Washington, D.C., May 28, 2008.

Other Sources

Abbott, Andrew, *The System of Professions: An Essay on the Division of Expert Labor*, Chicago: University of Chicago Press, 1988.

Bednarek, Mick, Thomas P. Odom, and Stephen Florich, "Expanding Jointness at the Joint Readiness Training Center," *Military Review*, January–February 2005, pp. 51–57.

Boese, R., P. Lewis, P. Frugoli, and K. Litwin, *Summary of O*NET 4.0 Content Model and Database*, Raleigh, N.C.: National Center for O*NET Development, 2001.

Bransford, John, *How People Learn: Brain, Mind, Experience and Schools*, Washington, D.C.: National Academy Press, 2000.

Brown, Fredric J., "Three Revolutions: From Training to Learning and Team Building," *Military Review*, July/August, 2003, pp. 54–61.

———, *Building High Performance Commander Leader Teams: Intensive Collaboration Enabled by Information Technology and Knowledge Management*, Alexandria, Va.: Institute for Defense Analyses, 2006.

Bryan, William Lowe, and Noble Harter, "Studies on the Telegraphic Language: The Acquisition of a Hierarchy of Habits," *The Psychological Review*, Vol. VI, No. 4, July 1899, pp. 345–375.

Bush, George, Executive Order 13434, "National Security Professional Development," May 17, 2007, *United States Federal Register*, Vol. 72, No. 98, pp. 28583–28585.

———, National Security Presidential Directive/NSPD-44, "Management of Interagency Efforts Concerning Reconstruction and Stabilization," December 7, 2005.

Chiarelli, Major General Peter, and Patrick Michaelis, "Winning the Peace: The Requirement for Full-Spectrum Operations," *Military Review*, July/August, 2005, pp. 4–17.

Doughty, Dr. Ralph, "Interagency Exchange Program: Preparing Leaders for the Future," PowerPoint briefing, June 23, 2008.

Ericsson, K. Anders, "The Influence of Experience and Deliberate Practice on the Development of Superior Expert Performance," in *The Cambridge Handbook of Expertise and Expert Performance*, Cambridge: Cambridge University Press, 2006a.

———, "An Introduction to the Cambridge Handbook of Expertise and Expert Performance: Its Development, Organization and Content," in *The Cambridge Handbook of Expertise and Expert Performance*, Cambridge: Cambridge University Press, 2006b.

Ericsson, K. Anders, Ralf Th. Krampe, and Clemens Tesch-Römer, "The Role of Deliberate Practice in the Acquisition of Expert Performance," *Psychological Review*, Vol. 100, No. 3, 1993, pp. 363–406.

Federal Emergency Management Agency (FEMA), "About the National Incident Management System (NIMS)." As of August 2010:
http://www.fema.gov/emergency/nims/AboutNIMS.shtm

Fern, E.F., "The Use of Focus Groups for Idea Generation: The Effects of Group Size, Acquaintanceship, and Moderator on Response Quantity and Quality," *Journal of Marketing Research*, Vol. 19, 1982, pp. 1–13.

Flavin, William, *Observations on Civil Military Operations During the First Year of Operation Enduring Freedom,* Carlisle Barracks, Pa.: Strategic Studies Institute, 2004.

France, Ministry of Defense, Etat Major des Armées (EMA), "Presentation du PC de Force," 2009. As of January 6, 2009:
http://www.defense.gouv.fr/ema/content/download/71265/660077/file/Pr%C3%A9sentation%20du%20PC%20de%20force.pdf

Geren, Pete, and George W. Casey, *2009 Army Posture Statement*, Washington, D.C.: Department of the Army, 2009.

Glenn, Russell W., and Gina Kingston, *Urban Battle Command in the 21st Century*, Santa Monica, Calif.: RAND Corporation, MG-181-A, 2005.
http://www.rand.org/pubs/monographs/MG181/

Glenn, Russell W., Christopher Paul, Todd C. Helmus, and Paul Steinberg, *"People Make the City," Executive Summary: Joint Urban Operations Observations from Afghanistan and Iraq*, Santa Monica, Calif.: RAND Corporation, MG-428/2-JFCOM, 2007.
http://www.rand.org/pubs/monographs/MG428.2/

Glenn, Russell W., Jody Jacobs, Brian Nichiporuk, Christopher Paul, Barbara Raymond, Randall Steeb, and Harry J. Thie, *Preparing for the Proven Inevitable: An Urban Operations Training Strategy for America's Joint Force*, Santa Monica, Calif.: RAND Corporation, MG-439-OSD/JFCOM, 2006.
http://www.rand.org/pubs/monographs/MG439/

Goerlitz, Walter, *History of the German General Staff: 1657–1945*, trans. Brian Battershaw, 10th Printing, New York: Praeger, 1959.

Goleman, Daniel, *Emotional Intelligence*, New York: Bantam Books, 1995.

Gonzales, Daniel, Michael Johnson, Jimmie McEver, Dennis Leedom, Gina Kingston, and Michael Tseng, *Network Centric Operations Case Study: The Stryker Brigade Combat Team*, Santa Monica, Calif.: RAND Corporation, MG-267-1-OSD, 2005.
http://www.rand.org/pubs/monographs/MG267-1/

Gott, Kenda D. (ed.), *Eyewitness to War: The U.S. Army in Operation AL FAJR, An Oral History*, Vols. I and II, Fort Leavenworth, Kan.: Combat Studies Institute, 2006.

Harrell, Margaret C., John F. Schank, Harry J. Thie, Clifford M. Graf II, and Paul Steinberg, *How Many Can Be Joint? Supporting Joint Duty Assignments,* Santa Monica, Calif.: RAND Corporation, MR-593-JS, 1996.
http://www.rand.org/pubs/monograph_reports/MR593/

Hirai, James T., and Kim L. Summers, "Leader Development and Education: Growing Leaders Now for the Future," *Military Review*, May/June 2005, pp. 85–96.

Holmes, J. Anthony, "Where Are the Civilians? How to Rebuild the U.S. Foreign Service," *Foreign Affairs*, Vol. 88, No. 1, January–February, 2009, pp. 148–160.

Howard, Michael, *The Franco-Prussian War*, London: Methuen and Company, 1981.

Joint Chiefs of Staff, *Unified Action Armed Forces*, JP 0-2, 2001.

———, *Interagency, Intergovernmental Organization, and Nongovernmental Organization Coordination During Joint Operations*, Vol. I, JP 3-08, 2006.

———, *Joint Operation Planning*, JP 5-0, 2006.

———, *Joint Operations*, JP 3-0, 2006.

———, *Joint Doctrine for the Armed Forces of the United States*, JP 1, 2007.

Joint Chiefs of Staff, Chairman, CJCSM 3122, *Joint Operations Planning and Execution System*, 2008.

Kirby, Sheila Nataraj, Al Crego, Harry J. Thie, Margaret C. Harrell, Kimberly Curry, and Michael S. Tseng, *Who Is "Joint"? New Evidence from the 2005 Joint Officer Management Census Survey*, Santa Monica, Calif.: RAND Corporation, TR-349-OSD, 2006.
http://www.rand.org/pubs/technical_reports/TR349/

Klein, Gary, *Sources of Power: How People Make Decisions*, Cambridge, Mass.: Massachusetts Institute of Technology Press, 1998.

Kleinmutz, Benjamin, "Why We Still Use Our Heads Instead of Formulas: Toward an Integrative Approach," *Psychological Bulletin*, Vol. 107, No. 3, pp. 296–310.

Krepinevich, Andrew F., *The Army and Vietnam*, Baltimore: Johns Hopkins University Press, 1988.

Leonard, Henry A., J. Michael Polich, Jeffrey D. Peterson, Ronald E. Sortor, and S. Craig Moore, *Something Old, Something New: Army Leader Development in a Dynamic Environment*, Santa Monica, Calif.: RAND Corporation, MG-281-A, 2006.
http://www.rand.org/pubs/monographs/MG281/

Lord, Robert G., and Karen J. Maher, "Cognitive Theory in Industrial and Organizational Psychology," in M.D. Dunnette and L.M. Hough (eds.), *Handbook of Industrial and Organizational Psychology*, Palo Alto, Calif.: Consulting Psychologists Press, 1991.

McCausland, Jeffrey D., *Developing Strategic Leaders for the 21st Century*, Carlisle Barracks, Pa.: Strategic Studies Institute, 2008.

Militello, Laura G., and Robert J.B. Hutton, "Applied Cognitive Task Analysis (ACTA): A Practitioner's Toolkit for Understanding Cognitive Task Demands," *Ergonomics,* Vol. 41, No. 11, 1998, pp. 1618–1641.

Morath, Ray, Christina Curnow, Candace Cronin, Arnold Leonard, and Tim McGonigle, *Identification of the Competencies Required of Joint Force Leaders*, Caliber Associates, 2007.

Morgan, David L. "Focus Groups," *Annual Review of Sociology*, Vol. 22, 1996, pp. 129–155.

Murdock, Clark A., Michelle A. Flournoy, Kurt M. Campbell, Pierre A. Chao, Julianne M. Smith, Anne A. Witkowski, and Christine E. Wormuth, *Beyond Goldwater-Nichols: Defense Reform for a New Strategic Era*, Phase II Report, Washington, D.C.: Center for Strategic and International Studies, 2005.

Murdock, Clark A., Michelle A. Flournoy, Christopher A. Williams, and Kurt M. Campbell, *Beyond Goldwater-Nichols: Defense Reform for a New Strategic Era*, Phase I Report, Washington, D.C.: Center for Strategic and International Studies, 2003.

National Center for O*NET Development, "O*NET OnLine." As of July 15, 2010:
http://online.onetcenter.org/

Neveux, General de Division Bruno, "L'Organisation du commandement pour l'opération ARTEMIS," *Doctrine*, No. 5, December 2004.

Norman, Geoff, Kevin Eva, Lee Brooks, and Stan Hamstra, "Expertise in Medicine and Surgery," in *The Cambridge Handbook of Expertise and Expert Performance*," Cambridge, U.K.: Cambridge University Press, 2006.

O*NET Resource Center, Abilities Questionnaire, 2003. As of January 5, 2009:
http://www.onetcenter.org/questionnaires.html

Perito, Robert M., *The U.S. Experience with Provincial Reconstruction Teams in Afghanistan: Lessons Identified,* United States Institute of Peace Special Report, Washington, D.C.: United States Institute of Peace, 2005.

Peterson, N. G., M. D. Mumford, W. C. Borman, P. R. Jeanneret, E. A. Fleishman, K. Y. Levin, et al., "Understanding Work Using the Occupational Information Network (O*NET): Implications for Practice and Research," *Personnel Psychology*, Vol. 54, 2001, pp. 451–492.

Phillips, Jennifer K., Gary Klein, and Winston R. Sieck, "Expertise and Decision Making: A Case for Training Intuitive Decision Making Skills," in D. J. Koehler and N. Harvey (eds.), *The Blackwell Handbook of Judgment and Decision Making*, Wiley and Blackwell, 2007.

Public Law 109-364, "John Warner National Defense Authorization Act for Fiscal Year 2007," 109th Cong., 2d Sess., October 17, 2006.

Rumsfeld, Donald, *Quadrennial Defense Review Report*, 2006.

Sackett, Paul R., and Roxanne M. Laczo, "Job and Work Analysis," in W. C. Borman, D. R. Ilgen, and R. J. Klimoski (eds.), *Handbook of Psychology, Volume 12: Industrial and Organizational Psychology*, Hoboken, N.J.: Wiley, 2003, pp. 21–37.

Schirmer, Peter, James C. Crowley, Nancy E. Blacker, Richard R. Brennan, Jr., Henry A. Leonard, J. Michael Polich, Jerry M. Sollinger, and Danielle M. Varda, *Leader Development in Army Units: Views from the Field*, Santa Monica, Calif.: RAND Corporation, MG-648-A, 2008.
http://www.rand.org/pubs/monographs/MG648/

Scott, Lynn M., Steve Drezner, Rachel Rue, and Jesse Reyes, *Compensating for Incomplete Domain Knowledge*, Santa Monica, Calif.: RAND Corporation, DB-517-AF, 2007.
http://www.rand.org/pubs/documented_briefings/DB517/

Shank, John F., Harry J. Thie, Jennifer Kawata, Margaret C. Harrell, Clifford M. Graf II, and Paul Steinberg, *Who Is Joint? Reevaluating the Joint Duty Assignment List,* Santa Monica, Calif.: RAND Corporation, MR-574-JS, 1996.
http://www.rand.org/pubs/monograph_reports/MR574/

Sonnentag, Sabine, "Using and Gaining Experience in Professional Software Development," in Eduardo Salas and Gary Klein (eds.), *Linking Expertise and Naturalistic Decision Making*, Mahwah, N.J.: Lawrence Erlbaum Associates, 2001.

Spirtas, Michael, Jennifer D.P. Moroney, Harry J. Thie, Joe Hogler, and Thomas-Durell Young, *Department of Defense Training for Operations with Interagency, Multinational, and Coalition Partners*, Santa Monica, Calif.: RAND Corporation, MG-707-OSD, 2008.
http://www.rand.org/pubs/monographs/MG707/

Thie, Harry J., Margaret C. Harrell, Roland J. Yardley, Marian Oshiro, Holly Ann Potter, Peter Schirmer, and Nelson Lim, *Framing a Strategic Approach for Joint Officer Management*, Santa Monica, Calif.: RAND Corporation, MG-306-OSD, 2005.
http://www.rand.org/pubs/monographs/MG306/

U.S. Army, *Career Management for Army Officers*, Technical Manual 20-605, 1948.

———, *The Army Universal Task List*, FM 7-15, 2003.

———, *Battle Focused Training*, FM 7-1, 2003.

———, *Mission Command: Command and Control of Army Forces*, FM 6-0, 2003.

———, *Army Planning and Orders Production*, FM 5-0, 2005.

———, *Army Leadership: Competent, Confident and Agile*, FM 6-22, 2006.

———, *Commissioned Officer Professional Development and Career Management*, Department of the Army Pamphlet 600-3, 2007.

———, *The Modular Force*, FMI 3-0.1, 2008.

———, *Operations*, FM 3-0, 2008.

———, *Stability Operations*, FM 3-07, 2008.

———, *Training for Full Spectrum Operations*, FM 7-0, 2008.

U.S. Army, Army Ground Forces, *Report of Army Ground Forces Study on Comparisons of General Officers and Colonels*, 1946.

U.S. Army, Center for Army Lessons Learned, *Catastrophic Disaster Response Staff Officers Handbook: Techniques and Procedures*, Fort Leavenworth, Kan.: Center for Army Lessons Learned, 2006.

U.S. Army, Training and Doctrine Command (TRADOC), TRADOC Pamphlet 525-3-7, *The U.S. Army Concept for the Human Dimension in Full Spectrum Operations, 2015–2024*, Fort Monroe, Va.: Headquarters, U.S. Army Training and Doctrine Command, 2008.

U.S. Department of Defense, Office of the Deputy Under Secretary of Defense for Personnel and Readiness, "Department of Defense Joint Officer Management System Fact Sheet," July 30, 2007.

U.S. Department of State, Office of the Coordinator for Reconstruction and Stabilization, *Post Conflict Reconstruction: Essential Tasks*, April 2005. As of October 2, 2009:
http://www.crs.state.gov/index.cfm?fuseaction=public.display&shortcut=J7R3

U.S. Government Accountability Office (GAO), *Contingency Contracting: DOD, State, and USAID Contracts and Contractor Personnel in Iraq and Afghanistan*, October 2008.

U.S. Joint Forces Command, *The Joint Operating Environment (JOE) 2008: Challenges and Implications for the Future Joint Force*, 2008.

U.S. Joint Knowledge Development and Distribution Capability, Joint Knowledge Online, *Certifications, Joint Qualified Officer (JQO) Points and Total Inventory Report*, January 18, 2008.

Wong, Leonard, *Developing Adaptive Leaders: The Crucible Experience of Operation Iraqi Freedom,* Carlisle Barracks, Pa.: Strategic Studies Institute, 2004.